Mathematics via Problems

PART 3: Combinatorics

MSRI Mathematical Circles Library

Mathematics via Problems

PART 3: Combinatorics

Mikhail B. Skopenkov & Alexey A. Zaslavsky

Translated from Russian by Sergei G. Shubin and Paul Zeitz

SIMONS LAUFER
MATHEMATICAL
SCIENCES INSTITUTE

Berkeley, California

AMS AMERICAN
MATHEMATICAL
SOCIETY

Providence, Rhode Island

This work was originally published in Russian by "МЦНМО" under the title Элементы математики в задачах, © 2018. The present translation was created under license for the American Mathematical Society and is published by permission.

This volume is published with the generous support of the Simons Foundation and Tom Leighton and Bonnie Berger Leighton.

2020 *Mathematics Subject Classification.* Primary 00-01, 00A07, 05-01, 52-01, 60-01, 94-01, 97K10, 97-01.

For additional information and updates on this book, visit
www.ams.org/bookpages/mcl-29

Library of Congress Cataloging-in-Publication Data

Part 2 was catalogued as follows:

Names: Zaslavskiĭ, Alekseĭ Aleksandrovich, 1960– author. | Skopenkov, Mikhail B., 1983– author.

Title: Mathematics via problems. Part 2. Geometry / Alexey A. Zaslavsky, Mikhail B. Skopenkov; translated by Sergei Shubin and Paul Zeitz.

Other titles: Élementy matematiki v zadachakh. English

Description: Berkeley, California: MSRI Mathematical Sciences Research Institute; Providence, Rhode Island American Mathematical Society, [2021] | Series: MSRI mathematical circles library, 1944-8074; 26 | Includes bibliographical references and index.

Identifiers: LCCN 2020057238 | ISBN 9781470448790 (paperback) | 9781470465216 (ebook)

Subjects: LCSH: Geometry–Study and teaching. | Problem solving. | Geometry–Problems, exercises, etc. | AMS: General – Instructional exposition (textbooks, tutorial papers, etc.). | General – General and miscellaneous specific topics – Problem books. | Geometry – Instructional exposition (textbooks, tutorial papers, etc.). | Convex and discrete geometry – Instructional exposition (textbooks, tutorial papers, etc.). | Algebraic geometry – Instructional exposition (textbooks, tutorial papers, etc.). | Mathematics education – Geometry – Comprehensive works. | Mathematics education – Instructional exposition (textbooks, tutorial papers, etc.).

Classification: LCC QA462 .Z37 2021 | DDC 516.0071–dc23

LC record available at https://lccn.loc.gov/2020057238

Part 3: ISBN: 9781470460105 (paperback), 9781470475895 (ebook).

Contents

Foreword

Problems, exercises, circles, and olympiads

This is a translation of Chapters 17–25 of the book *Mathematics Through Problems* by Mikhail B. Skopenkov and Alexey A. Zaslavsky and is part of the AMS/MSRI Mathematical Circles Library series. The goal of this series is to build a body of works in English that helps to spread the "Math Circle" culture.

A *mathematical circle* is an Eastern European notion. Math circles are similar to what most Americans would call a math club for kids, but with several important distinguishing features.

First, they are *vertically integrated*: young students may interact with older students, college students, graduate students, industrial mathematicians, professors, even world-class researchers, all in the same room. The circle is not so much a classroom as a gathering of young initiates with elder tribespeople, who pass down *folklore*.

Second, the "curriculum," such as it is, is dominated by *problems*, rather than specific mathematical topics. A problem, in contrast to an exercise, is a mathematical question that one doesn't know how, at least initially, to approach. For example, "What is 3 times 5?" is an exercise for most people, but it is a problem for a very young child. Computing 5^{34} is also an exercise, conceptually very much like the first, certainly harder, but only in a "technical" sense. And a question like "Evaluate $\int_2^7 e^{5x} \sin 3x\, dx$" is also an exercise—for calculus students—a matter of "merely" knowing the right algorithm and how to apply it.

Problems, by contrast, do not come with algorithms attached. By their very nature, they require *investigation*, which is both an art and a science, demanding technical skill along with focus, tenacity, and inventiveness. Math circles teach students these skills, not with formal instruction, but by *doing math* and observing others doing math. Students learn that a problem worth solving may require not minutes, but possibly hours, days, or even years of effort. They work on some of the classic folklore problems and discover how these problems can help them investigate other problems. They learn how not to give up and how to turn errors or failures into opportunities for more investigation. A child in a math circle learns to do exactly what a research mathematician does; indeed, he or she does independent research, albeit on a lower level, and often—although not always—on problems that others have already solved.

Finally, many math circles have a culture similar to a sports team, with intense camaraderie, respect for the "coach," and healthy competitiveness (managed wisely, ideally, by the leader/facilitator). The math circle culture is often complemented by a variety of problem-solving contests, often called *olympiads*. A mathematical olympiad problem is, first of all, a genuine problem (at least for the contestant) and usually requires an answer which is, ideally, a well-written argument (a "proof").

Why this book and how to use it

The Math Circles Library editorial board chose to translate Skopenkov and Zaslavsky's work from Russian into English because this book has an audacious goal—promised by its title—to develop mathematics through problems. This is not an original idea, nor just a Russian one. American universities have experimented for years with IBL (inquiry-based learning) and Moore-method courses, structured methods for teaching advanced mathematics through open-ended problem solving.[1]

But the authors' massive work is an attempt to curate sequences of problems for secondary students (their stated focus is high school students, but that can be broadly interpreted) that allow them to discover and recreate much of "elementary" mathematics (number theory, polynomials, inequalities, calculus, geometry, combinatorics, game theory, probability) and start edging into the sophisticated world of group theory, Galois theory, etc.

The book is impossible to read from cover to cover, nor should it be. Instead, the reader is invited to start working on problems that he or she finds appealing and challenging. Many of the problems have hints and solution sketches, but not all. No reader will solve all the problems. That's not the point—it is not a contest. Furthermore, some of the problems are not supposed to be solved but should be pondered. For example, Section 6 of Chapter 6 explores the unexpected connection between electrical circuits and random walks. In Chapter 7, the reader is encouraged to use similar ideas to analyze completely unrelated problems—dissecting squares into similar rectangles. Just because it is "too advanced" doesn't mean that it shouldn't be thought about!

Indeed, this is the philosophy of the book: mathematics is not a sequential discipline, where one is presented with a definition that leads to a lemma which leads to a theorem which leads to a proof. Instead it is an adventure, filled with exciting side trips as well as wild goose chases. The adventure is

---

[1]See, for example, https://en.wikipedia.org/wiki/Moore_method and http://www.jiblm.org.

its own reward, but it also, fortuitously, leads to a deep understanding and appreciation of mathematical ideas that cannot be accomplished by passive reading.

English-language references

Most of the references cited in this book are in Russian. However, there are many excellent books in English (some translated from Russian). Here is a very brief list, organized by topic. There are two bibliographies in this book. The references cited below are in the main bibliography at the end of the book.

Problem collections: *The USSR Olympiad Problem Book* [**SC**] is a classic collection of carefully discussed problems. Additionally, [**FKh**] and [**FBKYa1**, **FBKYa2**] are good collections of olympiads from Leningrad and Moscow, respectively. See also the collection of fairly elementary Hungarian contest problems [**Kur1**, **Kur2**, **Liu**] and the more advanced (undergraduate-level) Putnam Exam problems [**KKPV**].

Inequalities: See [**St**] for a comprehensive guide and [**AS**] for a more elementary text.

Geometry: *Geometry Revisited* [**CoxGr**] is a classic, and [**Chen**] is a more recent and very comprehensive guide to "olympiad geometry."

Polynomials and theory of equations: See [**B**] for an elementary guide and [**Bew**] for a historically motivated exposition of constructability and solvability and unsolvability.

Combinatorics: The best book in English, and possibly any language, is *Concrete Mathematics* [**GKP**].

Functions, limits, complex numbers, calculus: The classic book *Problems and Theorems in Analysis* by Pólya and Szegő [**PS**] is—like the current text—a curated selection of problems, but at a much higher mathematical level.

Paul Zeitz
April 2019

Introduction

What this book is about and whom it is for

A deep understanding of mathematics is useful both for mathematicians and for a high-tech professional.

This book is intended for high school students and undergraduates (in particular, those interested in Olympiads). For more details, see "Olympiads and mathematics" on p. xv. The book can be used both for self-study and for teaching. The references cited below are in the bibliography at the front of the book.

This book attempts to build a bridge (by showing that there is no gap) between ordinary high school exercises and the more sophisticated, intricate, and abstract concepts in mathematics. The focus is on engaging a wide audience of students to think creatively in applying techniques and strategies to problems motivated by "real world or real work" [**Mey**]. Students are encouraged to express their ideas, conjectures, and conclusions in writing. Our goal is to help students develop a host of new mathematical tools and strategies that will be useful beyond the classroom and in a number of disciplines [**IBL, Mey, RMP**].

The book contains the most standard "base" material (although we expect that at least some of this material is review, that not all is being learned for the first time). But the main content of the book is more complex material. Some topics are not well known in the traditions of mathematical circles but are useful both for mathematical education and for preparation for Olympiads.

The book is based on the classes taught by the authors at different times at the Independent University of Moscow, at various Moscow schools, in preparing the Russian team for the International Mathematical Olympiad, in the "Modern Mathematics" summer school, in the Kirov and Kostroma Summer Mathematical Schools, in the Moscow visiting Olympiad School, in the "Mathematical Seminar" and "Olympiad and Mathematics" circles, and also in the summer Conference of The Tournament of Towns.

Much of this book is accessible to high school students with a strong interest in mathematics.[1] We provide definitions for material that is not standard in the school curriculum, or provide references. However, many topics are difficult if you study them "from scratch." Thus, the ordering of the problems helps to provide "scaffolding." At the same time, many topics are *independent* of each other. For more details, see p. xvi, "How this book is organized".

Throughout this book we often use informal language when we explain how to apply combinatorics to "practical" problems. Most of the results can be easily made precise using the language of finite sets and operations with them. Such formalization becomes necessary only in Chapter 6, "Probability", where the lack of rigor would otherwise quickly lead to errors and paradoxes.

Learning by doing problems

We ascribe to the tradition of studying mathematics by solving and discussing problems. These problems are selected so that in the process of solving them the reader (more precisely, the solver) masters the fundamentals of important ideas, both classical and modern. The main ideas are developed incrementally with olympiad-style examples, in other words, by the simplest special cases, free from technical details. In this way, we show *how you can explore and discover these ideas on your own.*

Learning by solving problems is not just a serious approach to mathematics, but it also continues a venerable cultural tradition. For example, the novices in Zen monasteries study by reflecting on riddles ("koans") given to them by their mentors. (However, these riddles are rather more like paradoxes than what we consider to be problems.) See, for example, [**Su**]; compare with [**Pl**, pp. 26–33]. Here are some "math" examples: [**Ar04, BSh, GDI, KK08, Pr07-1, PoSe, SCY, Sk09, Va87-1, Zv**] which sometimes describe not only problems, but also the principles of selecting appropriate problems. For the American tradition, see [**IBL, Mey, RMP**].

Learning by solving problems is difficult, in part, because it generally does not create the *illusion* of understanding. However, the readers' efforts are fully rewarded by a deep understanding of the material, at first, with the ability to carry out similar (and sometimes rather different) reasoning. Eventually, while working on fascinating problems, the reader will be following the thought processes of the great mathematicians and may see how important concepts and theories naturally evolve. Hopefully this will help readers to make their own equally useful discoveries (not necessarily in math)!

---

[1]Some of the material is studied in some circles and summer schools by those who are just getting acquainted with mathematics (for example, 6th graders). However, the presentation is intended for a reader who already has at least a minimal mathematical culture. Younger students need a different approach; see, for example, [**GIF**].

Solving a problem, theoretically, requires only understanding its statement. Other facts and concepts are not needed. (However, useful facts and ideas will be developed when solving selected problems.) And you may need to know things from other parts of the book, as indicated in the instructions and hints. For the most important problems we provide hints, instructions, solutions, and answers, located at the end of each section. However, they should be referred to only after attempting the problems.

As a rule, we present the *formulation* of a beautiful or important result (in the form of problems) before its *proof*. In such cases, you may need to solve later problems in order to fully work out the proof. This is always explicitly mentioned in the text. Consequently, if you fail to solve a problem, please read on. This guideline is helpful because it simulates the typical research situation (see [**ZSS**, § 28]).

This book "is an attempt to demonstrate learning as *dialogue* based on solving and discussing problems" (see [**KK15**]).

Parting words *By A. Ya. Kanel–Belov*

To solve difficult Olympiad problems, at the very least you must have a robust knowledge of algebra (particularly algebraic transformations) and geometry. Most Olympiad problems (except for the easiest ones) require "mixed" approaches; rarely is a problem resolved by applying a method or idea in its pure form. Approaching such mixed problems involves combining several "crux" problems, each of which may involve single ideas in a "pure" form.

Learning to manipulate algebraic expressions is essential. The lack of this skill among Olympians often leads to ridiculous and annoying mistakes.

Olympiads and mathematics

> To him a thinking man's job was not to deny one reality
> at the expense of the other, but to include and to connect.
>
> U. K. Le Guin, *The Dispossessed*

Here are three common misconceptions: the best way to prepare for a math olympiad is by solving last year's problems; the best way to learn "serious" mathematics is by reading university textbooks; the best way to master any other skill is with no math at all. A further misconception is that it is difficult to achieve any two of these three goals simultaneously, because they are so divergent. The authors share the widespread belief that these three approaches miss the point and lead to harmful side effects: students either become too keen on emulation, or they study the *language* of mathematics rather than its *substance*, or they underestimate the value of robust math knowledge in other disciplines.

We believe that these three goals are not as divergent as they might seem. The foundation of mathematical education should be the *solution and discussion of problems interesting to the student, during which a student learns important mathematical facts and concepts.* This simultaneously prepares the students for math Olympiads and the "serious" study of mathematics, and it is good for their general development. Moreover, it is more effective for achieving success in any one of the three goals alone.

Research problems for high school students

Many talented high school and university students are interested in solving research problems. Such problems are usually formulated as complex questions broken into incremental steps; see, e.g., [**LKTG**]. The final result may even be unknown initially, appearing naturally only in the course of thinking about the problem. Working on such questions is useful in itself and is a good approximation to scientific research. Therefore it is useful if a teacher or a book can support and develop this interest.

For a description of successful examples of this activity, see, for example, projects in the Moscow Mathematical Conference of Schoolchildren [**M**]. While most of these projects are not completely original, sometimes they can lead to new results.

How this book is organized

You should not read each page in this book one after the other. You can choose a sequence of study that is convenient (or omit some topics altogether). Any section (or subsubsection) of the book can be used for a math circle session.

The book is divided into chapters and sections (some sections are divided into subsections), with a plan of organization outlined at the start of each section. If the material of another section is needed in a problem, then you can either ignore it or look up the reference. This gives greater freedom to the reader when studying the book, but at the same time it may require careful attentiveness.

The topics of each section are arranged approximately in order of increasing complexity. The numbers in parentheses after the item name indicate its "relative level": 1 is the simplest, and 4 is the most difficult. The first items (not marked with an asterisk) are basic; unless indicated otherwise, you should begin your study with them. The remaining ones (marked with an asterisk) can be returned to later; unless otherwise stated, they are independent of each other. As you read, try to *return* to old material, but at a new level. Thus you should end up studying different levels of a topic *not sequentially*, but as part of a mixture of topics.

The notation used throughout the book is given on p. xviii. Notations and conventions particular to a specific section are introduced at the beginning of that section. The book concludes with a subject index. The numbers in bold are the pages on which *formal definitions* of concepts are given.

Resources and literature

Besides sources for specialized material, we also tried to include the very best popular writing on the topics studied. We hope that this bibliography, at least as a first approximation, can guide readers through the sea of popular scientific literature in mathematics. However, the great size of this genre guarantees that many remarkable works were omitted. Please note that items in the bibliography are not necessary for solving problems in this book, unless explicitly stated otherwise.

Many of the problems are not original, but the source (even if it is known) is usually not specified. When a reference is provided, it comes after the statement of the problem, so that the readers can compare their solution with the one given there. When we know that many problems in a section come from one source, then we mentioned this.

We do not provide links to online versions of articles in the popular magazines *Quantum* and *Mathematical Enlightenment*; they can be found on the websites http://kvant.ras.ru, http://kvant.mccme.ru and http://www.mccme.ru/free-books/matpros.html.

Acknowledgements

We are grateful for the serious work of the translators of the book, Sergey Shubin and Paul Zeitz. We thank the reviewers for helpful comments, specifically, K. A. Knop (Chapter 5), D. V. Musatov (Section 5.5), and L. E. Mednikov (Chapters 1–7), and also the anonymous reviewers of selected materials. We thank A. I. Kanel-Belov, the author of a great body of work, who also contributed a number of useful ideas and comments. We thank our students for tricky questions and for pointing out errors. Further acknowledgments for specific items are given directly in the text

We apologize for any mistakes and will be grateful to readers for pointing them out.

Contact information

M. B. Skopenkov: King Abdullah University of Science and Technology, HSE University (Faculty of Mathematics), and the Institute for Information Transmission Problems of the Russian Academy of Sciences.

A. A. Zaslavsky: Central Economic Mathematical Institute, Moscow Power Energetic Institute, and Moscow School 1543.

Numbering and notation

The items inside each section are arranged approximately in order of increasing complexity of the material. The numbers in parentheses after the item name indicate its "relative level": 1 is the simplest, and 4 is the most difficult. The first items (not marked with an asterisk) are basic; unless indicated otherwise, you can begin to study the chapter with them. The rest of the items (marked with an asterisk) can be returned to later; unless otherwise stated, they are independent of each other.

Problem numbers are indicated in bold. If the statement of the problem involves creating the formulation of an assertion, then it is required to prove this assertion in the problem. More open-ended questions are called *challenges*; here you must come up with a clear wording and a proof; cf., for example, [**VIN**] found at the back of the book. In problems marked with the circle ° the reader is asked to provide just the answer without a proof.

The most difficult problems are marked with asterisks (*). If the statement of the problem asks you to "find" something, then you need to give a "closed form" answer (as opposed to, say, an unevaluated sum of many terms). Again, if you are unable to solve a problem, read on; later problems may turn out to be hints.

Notation

- $\lfloor x \rfloor = [x]$ — (lower) integer part of number x ("floor"), that is, the maximal integer not exceeding x.
- $\lceil x \rceil$ — the upper integer part of number x ("ceiling"), that is, the minimal integer, not less than x.
- $\{x\}$ — fractional part of number x, equal to $x - \lfloor x \rfloor$.

- $d \mid n$, or $n \vdots d$ — d *divides* n; that is, $d \neq 0$ and there exists such an integer k, such that $n = kd$ (the number d is called a *divisor* of the number n).

- $\mathbb{R}$, $\mathbb{Q}$, $\mathbb{Z}$, $\mathbb{N}$ — the sets of all real, rational, integer, and natural (that is, positive integer) numbers, respectively.

- $\mathbb{Z}_2$ — the set $\{0, 1\}$ of remainders upon division by 2 with the operations of addition and multiplication (modulo 2).

- $\mathbb{Z}_m$ — the set $\{0, 1, \ldots, m - 1\}$ of remainders upon division by m with the operations of addition and multiplication (modulo m). (Specialists in algebra often write this set as $\mathbb{Z}/m\mathbb{Z}$ and use $\mathbb{Z}_m$ for the set of *m-adic integers* for the prime m.)

- $\binom{n}{k}$ — the number of k-element subsets of n-element set (also denoted by C_n^k).

- $|X|$ — the number of elements in set X.

- $A - B = \{x \mid x \in A \text{ and } x \notin B\}$ — the difference of the sets A and B.

- $A \sqcup B$ — the disjoint union of the sets A and B, that is, the union of the sets A and B, if they have no common elements.

- $A \subset B$ — means the set A is contained in the set B. In some books, this is denoted by $A \subseteq B$, and $A \subset B$ means "the set A is in the set B and is not equal to B."
- We abbreviate the phrase "Define x by a" with $x := a$.

Bibliography

[GDI]   A. A. Glibichuk, A. B. Dajnyak, D. G. Ilyinsky, A. B. Kupavsky, A. M. Rajgorod-sky, A. B. Skopenkov, and A. A. Chernov, *Elements of discrete mathematics in problems*, MCCME, Moscow, 2016 (Russian).
Abridged version: http://www.mccme.ru/circles/oim/discrbook.pdf.

[GIF]   S. A. Genkin, I. V. Itenberg, and D. V. Fomin, *Leningrad mathematical circles*, Kirov, 1994 (Russian).

[GKP]   Ronald L. Graham, Donald E. Knuth, and Oren Patashnik, *Concrete mathematics: A foundation for computer science*, 2nd ed., Addison-Wesley Publishing Company, Reading, MA, 1994. MR1397498

[VIN]   O. Ya. Viro, O. A. Ivanov, N. Yu. Netsvetaev, and V. M. Kharlamov, *Elementary topology: problem textbook*, American Mathematical Society, Providence, RI, 2008, DOI 10.1090/mbk/054. MR2444949

[ZSS]   A. Zaslavsky, A. Skopenkov and M. Skopenkov (eds.), *Elements of mathematics in problems: through olympiads and circles to the profession*, MCCME, Moscow, 2017 (Russian). Abridged version: http://www.mccme.ru/circles/oim/materials/sturm.pdf.

[Ar04]   Vladimir Arnold, *Problems for children 5 to 15 years old*, Eur. Math. Soc. Newsl. **98** (2015), 14–20. Excerpt from *Lectures and problems: a gift to young mathematicians*, American Mathematical Society, Providence, RI, 2015 [MR3409220]. MR3445185

[BSh]   A. D. Blinkov and A. V. Shapovalov (eds.), Book series "School math circles".

[IBL]   http://en.wikipedia.org/wiki/Inquiry-based_learning

[KK08]   A. Ya. Kanel-Belov and A. K. Kovalji, *How to solve non-standard problems*, MCCME, Moscow, 2008 (Russian).

[KK15]   A. K. Kovalji and A. Ya. Kanel-Belov, *Mathematics Classes — Leaflets and Dialogue*, Mathematical Education **3** (2015), no. 19, 206–233 (Russian).

[LKTG]   Summer Conferences of the Tournament of Towns, http://www.turgor.ru/en/lktg/index.php.

[M]   Moscow Mathematical Conference of High School Students, http://www.mccme.ru/mmks/index.htm.

[Mey]   D. Meyer, Starter Pack, http://blog.mrmeyer.com/starter-pack.

[PoSe]   G. Pólya and G. Szegő, *Problems and theorems in analysis. Vol. I: Series, integral calculus, theory of functions*, translated from the German by D. Aeppli, Die Grundlehren der mathematischen Wissenschaften, Band 193, Springer-Verlag, New York-Berlin, 1972. MR0344042

[Pr07-1] V. V. Prasolov, *Problems in Plane Geometry*, MCCME, Moscow, 2007 (Russian).

[Pl]   Plato, *Phaedo*, Kindle edition, Amazon Digital Services, 2012.

[Ro04]   V. A. Rokhlin, *Lecture on teaching mathematics to nonmathematicians*, Mathematical Education **3** (2004), no. 8, 21–36 (Russian).

[RMP]   Ross Mathematics Program, http://u.osu.edu/rossmath.

[SCY]   D. O. Shklarsky, N. N. Chentzov, and I. M. Yaglom, *The USSR olympiad problem book: selected problems and theorems of elementary mathematics*, translated from the third Russian edition by John Maykovich, revised and edited by Irving Sussman, reprint of the 1962 translation, Dover Publications, Inc., New York, 1993. MR1252851

[Sk09] A. B. Skopenkov, *Fundamentals of differential geometry in interesting problems*, MCCME, Moscow, 2009 (Russian). arXiv:0801.1568.

[Su]   D. T. Suzuki, *An introduction to Zen Buddhism*, Kindle edition, Amazon Digital Services, 2018.

[Va87-1] N. B. Vasiliev et al., *Mathematical olympiads by correspondence*, Nauka, Moscow 1987 (Russian).

[Zv]   A. K. Zvonkin, *Math from three to seven: the story of a mathematical circle for preschoolers*, American Mathematical Society, Providence, RI, 2012.

Chapter 1

Counting

This introductory chapter focuses on the question, "How many objects are there with given properties?" For further study, we recommend Chapter 1 of [**GDI**].

1. How many ways? (1)
By A. A. Gavrilyuk and D. A. Permyakov

1.1.1. (a) Call a positive integer *nice* if it contains only even digits. Write down all the nice two-digit numbers; how many are there?

(b) How many five-digit numbers are nice?

(c) How many six-digit numbers have at least one even digit?

(d) Which are there more of: seven-digit numbers that contain a 1 or seven-digit numbers that have no 1's?

1.1.2. We wish to form a committee of eight people, chosen from two mathematicians and ten economists. In how many ways can this be done if at least one mathematician is to be included on the committee?

1.1.3. (a) Find the sum of all seven-digit numbers that can be obtained by permuting the digits $1, \ldots, 7$.

(b) Find the sum of all four-digit numbers with no zeros and no repeating digits.

(c) Find the sum of all four-digit numbers that do not contain repeating digits.

1.1.4. (a) Black and white kings occupy two squares of a chessboard. The player moves a king (alternating between black and white king) to an adjacent square (horizontally, vertically, or diagonally). The kings are friends, so they can occupy neighboring squares but not be in the the same square. Is it possible to get the kings to occupy every pair of squares (s_1, s_2) with the white king on s_1 and the black king on s_2, exactly once?

(b) The same question, but now the kings are unable to move diagonally.

1.1.5. (a) Find the sum of all six-digit numbers obtained by all permutations of the digits $4, 5, 5, 6, 6, 6$.

(b) Find the sum of all ten-digit numbers obtained by all permutations of the digits $4, 5, 5, 6, 6, 6, 7, 7, 7, 7$.

1.1.6. (a) Tom Sawyer was commissioned to paint 8 boards of a fence white. He is lazy, so he will paint no more than 3 boards. In how many ways can he do this?

(b) How many ways are there to paint no more than 5 boards?

(c) How many ways are there to paint any number of boards?

Suggestions, solutions, and answers

1.1.1. *Answers*: (b) 2500;   (c) 884 375;   (d) the numbers containing a 1.

(b) *Solution* (A. Kolochenkov). The first digit of a nice number can be 2, 4, 6, or 8, so there are just 4 options. For each digit in the second to fifth place, there are 5 options: 0, 2, 4, 6, 8. So the total amount of nice numbers is $4 \cdot 5 \cdot 5 \cdot 5 \cdot 5 = 2500$.

This reasoning is called the *product rule* in combinatorics and is discussed in detail in [**Vil71b**].

(c) *Solution* (A. Kolochenkov). Subtract from the total number of six-digit numbers the number of six-digit numbers consisting entirely of odd digits. Then there will remain numbers in which at least one digit is even. Since there are only five odd digits, the product rule yields $9 \cdot 10 \cdot 10 \cdot 10 \cdot 10 \cdot 10 - 5 \cdot 5 \cdot 5 \cdot 5 \cdot 5 \cdot 5 = 884\,375$.

This reasoning (the total quantity of numbers is equal to the sum of the quantity of numbers from odd digits and the quantity of numbers with even digit) is called the *sum rule* and is discussed in detail in article [**Vil71b**].

1.1.2. *Answer*: 450.

1.1.3. *Answers*: (a) $22\,399\,997\,760$; (b) $16\,798\,320$; (c) $24\,917\,760$.

(a) *Solution* (T. Cherganov). There are 7! permutations of seven digits. Each digit will occur in all positions the same number times, equal to $\frac{7!}{7} = 6!$. Then the sum of the digits in each position is equal to $6! \cdot (1 + 2 + \cdots + 7) = 20160$. So the sum of all numbers is equal to

$$20160 + 20160 \cdot 10 + 20160 \cdot 100 + \cdots + 20160 \cdot 1000000$$
$$= 20160 \cdot 1111111 = 22399997760.$$

(b) *Solution* (S. Kudrya). We calculate the sum of the digits in each position. Each digit is included in the sum $8 \cdot 7 \cdot 6$ times; indeed, for the first of the remaining positions there are 8 possibilities to place a digit, for the second one there are 7, for the third there are 6. Therefore, the amount in each position is $8 \cdot 7 \cdot 6 \cdot 1 + 8 \cdot 7 \cdot 6 \cdot 2 + 8 \cdot 7 \cdot 6 \cdot 3 + \cdots + 8 \cdot 7 \cdot 6 \cdot 9 = 8 \cdot 7 \cdot 6 \cdot (1 + 2 + 3 + \cdots + 9) = 8 \cdot 7 \cdot 6 \cdot \frac{9 \cdot 10}{2} = 15120$. Multiplying this by 1111 produces the desired sum of numbers.

1.1.4. (a) *Answer*: yes. *Hint*: explicitly construct an example.

(b) *Answer*: no. *Hint*: count the number of positions.

Suggestion. Suppose they can. The number of "one-color" positions for which the kings stand on the cells of the same color is $64 \cdot 31$. The number of "multi-colored" positions is $64 \cdot 32$. The type of position flips after each move of a king (they cannot move diagonally). If all these positions appear once, then quantities $64 \cdot 32$ and $64 \cdot 31$ differ by no more than one. Contradiction.

1.1.5. *Answers*: (a) $35\,555\,520$; (b) $83\,999\,999\,991\,600$.

(a) Six digits can be arranged in a row in $6! = 720$ ways. However, since the sixes are indistinguishable and the fives are indistinguishable, the total number of distinguishable rearrangements of these six digits is $\frac{6!}{3! \cdot 2!} = 60$. Each digit occurs an equal number of times in each of the six positions. The four occurs 60 times, which means that in each position it occurs 10 times. Five occurs 120 times, which means that in each position it occurs 20 times. The six occurs 180 times, which means that in each position it occurs 30 times. The sum of the digits in each position is $4 \cdot 10 + 5 \cdot 20 + 6 \cdot 30 = 320$. So the sum of all numbers is $320 + 320 \cdot 10 + 320 \cdot 100 + 320 \cdot 1000 + 320 \cdot 10\,000 + 320 \cdot 100\,000 = 320 \cdot 111\,111$.

(b) *First solution* (V. Tsepelev). We will count the number of occurrences of each of the digits in a specific position. Having fixed one of the digits, we may freely arrange the digits in the remaining nine. Note that 5 occurs twice, 6 three times, and 7 four times. Permutations of identical digits give identical numbers, so 4 will occur $\frac{9!}{2!3!4!}$ times. Similarly, 5 will occur $\frac{9!}{1!3!4!}$ times, and 6 and 7 will occur $\frac{9!}{1!2!2!4!}$ and $\frac{9!}{1!2!3!3!}$ times.

Now we compute the sum of the digits in each position: $4 \cdot \frac{9!}{2!3!4!} + 5 \cdot \frac{9!}{3!4!} + 6 \cdot \frac{9!}{2!2!4!} + 7 \cdot \frac{9!}{2!3!3!} = 75\,600$. Finally, multiply by $111\,111\,111$.

Second solution (V. Tsepelev). The answer can be obtained faster if you notice that the average of all digits of the set $4, 5, 5, 6, 6, 6, 7, 7, 7, 7$ is equal to 6. Therefore, the sum will be the same as for the same quantity of repeats of $6\,666\,666\,666$. The total number of repeats is $\frac{10!}{1!2!3!4!}$, so the sum is $6\,666\,666\,666 \cdot \frac{10!}{1!2!3!4!} = 83\,999\,999\,991\,600$.

1.1.6. (a) *Answer*: 93.

Suggestion. If Tom paints no more than three boards, then he can paint $0, 1, 2$, or 3 boards. He can do it in $\binom{8}{0} + \binom{8}{1} + \binom{8}{2} + \binom{8}{3} = 93$ ways.

(b) *Answer*: 219.

Similarly, no more than 5 boards can be painted in $\binom{8}{0} + \binom{8}{1} + \binom{8}{2} + \binom{8}{3} + \binom{8}{4} + \binom{8}{5} = 219$ ways.

(c) *Answer*: $256 \left(= 2^8 = \binom{8}{0} + \binom{8}{1} + \binom{8}{2} + \binom{8}{3} + \binom{8}{4} + \binom{8}{5} + \binom{8}{6} + \binom{8}{7} + \binom{8}{8} \right)$.

2. Sets of subsets (2) *By D. A. Permyakov*

1.2.1. Every evening, Uncle Chernomor selects 9 or 10 bogatyrs (hero warriors) from his company of 33 for guard duty. What are the fewest number of evenings such that it is possible that each of the bogatyrs went on duty for the same number of times?

1.2.2. A classroom has 33 students. Each student was asked how many students there are with the same first name as his in the class and how many students there are with the same last name as his (including relatives). It turned out that all integers from 0 to 10, inclusive, were reported. Must the classroom contain two students with the same first and last names?

1.2.3. (a) What is the maximum number of pairwise intersecting subsets that can be chosen from a set of 100 elements?

(b) In how many ways can a set of n elements be decomposed into 2 subsets?

(c) How many different unordered pairs of disjoint subsets can be chosen from a set of n elements?

1.2.4. One is given 2007 sets, each of which contains 40 elements. Any two of these sets have exactly one common element. Is there necessarily an element belonging to each of these sets?

1.2.5. In the set consisting of 100 elements, choose 101 distinct three-element subsets. Must there exist two subsets among them that have exactly one common element?

1.2.6. A questionnaire was conducted in a country in which the respondent was required to name his favorite writer, artist, and composer. It turned out that each writer, artist, or composer was mentioned as a favorite by no more than k people. Is it true that all of the surveyed people can always be divided into no more than $3k - 2$ groups, so that in each group any two people have completely different tastes?

1.2.7. Consider all nonempty subsets of the set $\{1, 2, \ldots, N\}$ that do not contain consecutive numbers. For each subset, calculate the product of its elements. Find the sum of the squares of these products.

Suggestions, solutions, and answers

1.2.1. *Answer*: 7.

Suggestion. Let 9 warriors be on duty for $m \geq 0$ days, and let 10 warriors be on duty for $n \geq 0$ days. Then (since each of the warriors went on duty the same number of times) the number $9m + 10n$ must be divisible by 33. Case by case analysis of all possible values of n shows that $9m + 10n = 33$ has no solutions in nonnegative integers. However, $9m + 10n = 66$ has the solution $m = 4$, $n = 3$. It is easy to construct an example spanning $m + n = 7$

days where each warrior serves exactly 2 times. If $9m + 10n \geq 99$, then $m + n \geq \frac{99}{10} > 7$. Therefore, the minimum number of days is 7.

1.2.2. *Answer*: yes.

Solution (T. Cherganov). Divide the students into groups by first and last name. Then each student will fall into exactly two groups. There are at least 11 such groups, since all numbers from 0 to 10 inclusive were reported. The total number of people in the 11 groups is not less than $\frac{1+2+\cdots+11}{2} = 33$. This means that there are exactly 11 groups, and the number of people in them is $1, 2, \ldots, 11$, respectively. Consider the group of 11 students. Without loss of generality, assume that 11 people have the same last name. Each of them is a member of one of the remaining 10 groups (by first name). By the pigeonhole principle, at least one of these groups (by first name) includes 2 people that have the same last name.

1.2.3. (a) *Answer*: 2^{99}.

Suggestion. The set has 2^{100} subsets. Divide them in pairs: put each subset in a pair with its complement. Then there will be a total of 2^{99} pairs, and from each pair we can choose no more than one subset. To find 2^{99} pairwise intersecting subsets, just take all subsets containing some fixed element.

(b) *Answer*: 2^{n-1}.

Consider the partition of the set into two subsets to be the choice of some subset and its complement. Choosing a subset, we can either select or not select a particular element. So the number of ways to choose a subset is 2^n. But we counted each partition twice, since we can reverse the role of "subset" and "complement". Therefore, there are 2^{n-1} different partitions.

(c) *Answer*: $\frac{3^n - 1}{2}$.

Suggestion. First calculate the number of ways to select an ordered pair of disjoint subsets. There are three options for each element: it either lies in the first subset, the second one, or neither subset. There will be 3^n such ordered pairs. An unordered pair of two empty subsets corresponds to one ordered pair of empty subsets. Any other unordered pair corresponds to two ordered pairs of subsets. This means that a total number of unordered pairs is $\frac{3^n - 1}{2}$.

1.2.4. *Answer*: yes.

Solution (V. Tsepelev). We denote given sets $J_1, \ldots, J_{2007}$. Let $J_1 = (a_1, \ldots, a_{40})$. Split the numbers corresponding to the indices in the family $J_2, \ldots, J_{2007}$ into sets of the form $X_i = \{j \geq 2 : a_i \in J_j\}$, $i = 1, \ldots, 40$, since each pair of sets has exactly one common element.

Since $\frac{2007-1}{40} = 50.15$, one of X_i—without loss of generality, suppose it is X_1—contains at least 51 elements. Consider two cases. If $X_1 = \{2, \ldots, 2007\}$, then the statement is proved (a_1 belongs to all sets; there are no other common elements). If not, then assume without loss of generality that $2007 \notin X_1$ and that $2, \ldots, 52 \in X_1$. Then for all $j = 2, \ldots, 52$ the intersection $X_j \cap X_{2007}$ will contain exactly one element, and this element is different for different indices j (since any two sets have exactly one common element). But this is not possible because X_{2007} consists of only 40 elements, which is less than 51.

1.2.5. *Answer*: yes.

1.2.6. *Answer*: yes.

1.2.7. *Answer*: $(N+1)! - 1$.

3. The principle of inclusion-exclusion (2)
By D. A. Permyakov

This section is devoted to the proof and application of the inclusion-exclusion formula, also known as the principle of inclusion-exclusion. It allows you to answer the question, "How many objects are there with given properties?" in many difficult cases. It will require basic skills for solving combinatorics problems. In particular, one must be able to give rigorous proofs using one-to-one correspondences and the rules of sums and products. For example, it is useful to solve problems from Section 1 of this chapter or problems from article [**Vil71b**].

1.3.1. In how many ways can you rearrange the numbers from 1 to n so that

(a) neither 1 nor 2 occurred in its original position;

(b) exactly one of the numbers 1, 2, and 3 stayed in its original position;

(c) none of the numbers 1, 2, and 3 occurred in their original positions;

(d) none of the numbers 1, 2, 3, and 4 occurred in its original position?

Define the Euler function $\varphi(n)$ to be the number of integers between 1 and n relatively prime to the number n.

1.3.2. (a) Find the number of integers from 1 to 1001 not divisible by any of the numbers 7, 11, 13.

(b) Find $\varphi(1)$, $\varphi(p)$, $\varphi(p^2)$, $\varphi(p^\alpha)$, where p is a prime, $\alpha > 2$.

(c) Prove that $\varphi(n) = n\left(1 - \frac{1}{p_1}\right) \ldots \left(1 - \frac{1}{p_s}\right)$, where $n = p_1^{\alpha_1} \cdot \cdots \cdot p_s^{\alpha_s}$ is the canonical decomposition of the number n into distinct primes p_k.

1.3.3. (a) On the floor of a room with area $24\,\mathrm{m}^2$ there are three area rugs (of arbitrary shape) each with area $12\,\mathrm{m}^2$. Then there exist two rugs with the area of the intersection at least $4\,\mathrm{m}^2$.

(b) There are five patches (of arbitrary shape) on a caftan.[1] The area of each patch is more than three-fifths of the area of caftan. Then there are two patches such that the area of their intersection is more than one-fifth the area of the caftan.

(c)* Same as in part (b), but the area of each patch is assumed to be greater than *half* the area of the caftan.

---

[1] A *caftan* is an old-fashioned Russian cloth similar to a jacket.

In this section, we encounter problems of the following type: a finite set U and a set of properties (subsets) $A_k \subset U$, $k = 1, \dots, n$, are given. We wish to find the number of elements for which at least one of the A_k properties is satisfied (i.e., $|A_1 \cup \cdots \cup A_n|$), or the number of elements for which none of the properties A_k is satisfied (i.e., $|U - (A_1 \cup \cdots \cup A_n)|$). For this, two versions of the inclusion-exclusion principle are used (see problem 1.3.5(b)). Moreover, if in all intersections of the sets of the family the number of elements depends only on the number of intersected sets, then the formula can be simplified (see problem 1.3.5(a)).

1.3.4. Consider the subsets A_1, A_2, A_3, A_4 of a finite set U. Prove the following equalities:

(a) $A_1 \cup A_2 = (A_1 \backslash A_2) \sqcup (A_1 \cap A_2) \sqcup (A_2 \backslash A_1)$;

(b) $|A_1 \cup A_2| = |A_1| + |A_2| - |A_1 \cap A_2|$;

(c) $|A_1 \cup A_2 \cup A_3| = |A_1| + |A_2| + |A_3| - |A_1 \cap A_2| - |A_2 \cap A_3| - |A_1 \cap A_3| + |A_1 \cap A_2 \cap A_3|$.

(d) The number of elements in U that do not belong to any of the subsets A_1, A_2, A_3 is

$$|U| - |A_1| - |A_2| - |A_3| + |A_1 \cap A_2| + |A_2 \cap A_3| + |A_1 \cap A_3| - |A_1 \cap A_2 \cap A_3|.$$

(e) For $k = 1, 2, 3, 4$ define

$$M_k := \sum_{1 \le i_1 < \cdots < i_k \le 4} |A_{i_1} \cap A_{i_2} \cap \cdots \cap A_{i_k}|.$$

Prove that the number of elements in A that do not belong to any of A_i is equal to $|U| - M_1 + M_2 - M_3 + M_4$.

(f) Using the above notation, the number of elements belonging to exactly one of the sets A_i is $M_1 - 2M_2 + 3M_3 - 4M_4$.

1.3.5. Inclusion-exclusion principle. Consider the subsets $A_1, \dots, A_n$ of a finite set U. By definition set $\bigcap\limits_{j \in \emptyset} A_j := U$.

(a) Suppose that the number $\alpha_{|S|} := \left| \bigcap\limits_{j \in S} A_j \right|$ depends only on the size $|S|$ of the set $S \subset \{1, \dots, n\}$ of indices, but not on the set itself. Then

$$|A_1 \cup \cdots \cup A_n| = \sum_{k=1}^{n} (-1)^{k+1} \binom{n}{k} \alpha_k,$$

$$|U - (A_1 \cup \cdots \cup A_n)| = \sum_{k=0}^{n} (-1)^{k} \binom{n}{k} \alpha_k.$$

(b) Define $M_k := \sum_{S \in \binom{n}{k}} \left| \bigcap_{j \in S} A_j \right|$, where the indicated summation is over all k-element subsets of the set $\{1, \ldots, n\}$. In particular, $M_0 := |U|$. Then

$$|A_1 \cup \cdots \cup A_n| = M_1 - M_2 + M_3 - \cdots + (-1)^{n+1} M_n,$$
$$|U - (A_1 \cup \cdots \cup A_n)| = M_0 - M_1 + M_2 + \cdots + (-1)^n M_n.$$

(c) **Bonferroni Inequalities.** For any $0 \le s < n/2$, the following inequalities hold:

$$M_1 - M_2 + M_3 - \cdots - M_{2s} \le |A_1 \cup \cdots \cup A_n|$$
$$\le M_1 - M_2 + M_3 - \cdots + M_{2s+1},$$
$$M_0 - M_1 + M_2 - \cdots + M_{2s} \ge |U - (A_1 \cup \cdots \cup A_n)|$$
$$\ge M_0 - M_1 + M_2 - \cdots - M_{2s+1}.$$

(d) The number of elements belonging to exactly r from the subsets $A_1, \ldots, A_n$ is equal to $\sum_{k=r}^{n} (-1)^{k-r} \binom{k}{r} M_k$.

1.3.6. A shelf holds 10 different books.

(a) In how many ways can they be rearranged so that none of the books stays in place?

(b) Prove that the number of rearrangements for which exactly 4 books stay in place is more than $50\,000$.

In the next problem, the answer can be in terms of sums (similar to the inclusion-exclusion principle).

1.3.7. (a) In how many ways can you arrange 20 tourists into 5 different houses so that not a single house is empty?

(b) How many different surjections $f \colon \mathbb{Z}_k \to \mathbb{Z}_n$ are there?

1.3.8. The numbers $1, 2, \ldots, n$ are placed on a circle. Find the number of ways to select k of them so that no two selected numbers were adjacent.

(b) Find the number of ways to seat n pairs of warring knights at a round table with numbered seats so that no two warring knights sit next to one another.

1.3.9. A cube with edge of length 20 is divided into 8000 unit cubes, and each unit cube is given a numerical label. It is known that in each row of 20 cubes parallel to an edge of the cube, the sum of the labels is 1 (rows in all three directions are considered). The label 10 is used on at least one cube. Three layers of $1 \times 20 \times 20$ pass through this cube parallel to the faces of the cube. Find the sum of all the labels not contained in any of these layers.

1.3.10.* How many six-digit numbers are there with no two 7's adjacent and with

(a) no more than three 7's;
(b) not more than four 7's;
(c) any number of 7's?

1.3.11.* Prove the following formula:

$$n! \cdot x_1 x_2 \ldots x_n = (x_1 + x_2 + \cdots + x_n)^n$$
$$- \sum_{1 \le i_1 < i_2 < \cdots < i_{n-1} \le n} (x_{i_1} + x_{i_2} + \cdots + x_{i_{n-1}})^n$$

$$+ \sum_{1 \le i_1 < i_2 < \cdots < i_{n-2} \le n} (x_{i_1} + x_{i_2} + \cdots + x_{i_{n-2}})^n - \cdots + (-1)^{n-1} \sum_{i=1}^{n} x_i^n.$$

Suggestions, solutions, and answers

1.3.1. (a)*Answer*: $n! - 2(n-1)! + (n-2)!$.

Suggestion. There are $n!$ ways in total to rearrange our numbers. Subtract from them the $(n-1)!$ permutations for which the number 1 remains in place. Subtract also the $(n-1)!$ permutations for which the number 2 remains in place. However, there are $(n-2)!$ permutations for which both numbers 1 and 2 remain in place, which we "subtracted twice". Thus, the number $(n-2)!$ needs to be added, yielding $n! - 2(n-1)! + (n-2)!$.

Comment. This solution is formalized by the principle of inclusion-exclusion, problem 1.3.5, which we assume you have proved and will use in subsequent problems.

1.3.2. (a)*Answer*: 720.

First suggestion. For each divisor j of 1001, let A_j denote the set of numbers from 1 to 1001 divisible by j. Then

$$|A_j| = \frac{1001}{j} \quad \text{and} \quad A_{p_1} \cap \cdots \cap A_{p_k} = A_{p_1 \cdots p_k}$$

for distinct primes $p_1, \ldots, p_k$. Therefore, we wish to compute

$$|\{1, \ldots, 1001\} - (A_7 \cup A_{11} \cup A_{13})|$$
$$= 1001 - |A_7| - |A_{11}| - |A_{13}| + |A_7 \cap A_{11}| + |A_7 \cap A_{13}|$$
$$+ |A_{11} \cap A_{13}| - |A_7 \cap A_{11} \cap A_{13}|$$
$$= 1001 - |A_7| - |A_{11}| - |A_{13}| + |A_{77}| + |A_{91}| + |A_{143}| - |A_{1001}|$$
$$= 1001 - 143 - 91 - 77 + 7 + 11 + 13 - 1 = 720.$$

Second suggestion. Use the equality $\varphi(1001) = \varphi(7)\varphi(11)\varphi(13)$.

(c) *Suggestion.* For any $j \mid n$, define A_j as a subset of the set $\{1, \ldots, n\}$, consisting of numbers that are divisible by j. Clearly

$$|A_j| = \frac{n}{j} \quad \text{and} \quad A_{p_1} \cap \cdots \cap A_{p_k} = A_{p_1 \cdots p_k}$$

for distinct primes $p_1, \ldots, p_k$.

Define

$$M_k := \sum_{1 \leq j_1 < \cdots < j_k \leq s} \frac{n}{p_{j_1} p_{j_2} \cdots \cdots p_{j_k}} \quad \text{for} \quad k \geq 1 \quad \text{and} \quad M_0 := n.$$

By definition, $\varphi(n) = |\{1, \ldots, n\} - (A_{p_1} \cup \cdots \cup A_{p_s})|$, whence the inclusion-exclusion principle (1.3.5(b)) yields

$$\varphi(n) = M_0 - M_1 + M_2 - \cdots + (-1)^s M_s = n\left(1 - \frac{1}{p_1}\right) \cdots \left(1 - \frac{1}{p_s}\right).$$

1.3.3. (a) Let A_j denote the set of points covered by the jth rug, and let $|A_j|$ be the area of the jth rug.

Suppose the statement is false; i.e., for any distinct j, k we have $|A_j \cap A_k| < 4$. Consequently,

$$\begin{aligned}
24 &\geq |A_1 \cup A_2 \cup A_3| \\
&\geq |A_1| + |A_2| + |A_3| - |A_1 \cap A_2| - |A_2 \cap A_3| - |A_1 \cap A_3| \\
&> 12 + 12 + 12 - 4 - 4 - 4 = 24,
\end{aligned}$$

where the second inequality follows from the analogue of the Bonferroni inequality for areas (1.3.5(c)), which produces a contradiction.

(b), (c) These problems are discussed in detail in the article [**Yag74**].

1.3.6. (a) *Answer*: $1\,334\,961 = \text{round}(10!/e)$, where

$$e := 1 + \frac{1}{1!} + \frac{1}{2!} + \frac{1}{3!} + \cdots + \frac{1}{n!} + \cdots$$

and $\text{round}(x)$ is the nearest integer to a real number x.

Suggestion. Set $n := 10$. Let U be the set of all book permutations, and let A_j be the set of book permutations for which the jth book remains in its original place. Consider an arbitrary k-element subset $S \subset \{1, \ldots, n\}$. Then $\bigcap_{j \in S} A_j$ consists of those permutations of books for which each of the books $j \in S$ remains in its original place. Therefore

$$\left| \bigcap_{j \in S} A_j \right| = (n - k)(n - k - 1) \ldots 1 = (n - k)!.$$

Using the principle of inclusion-exclusion, we obtain

$$|U - (A_1 \cup \cdots \cup A_n)| = \sum_{k=0}^{n} (-1)^k \binom{n}{k}(n - k)!$$

$$= n! - \frac{n!}{1!} + \frac{n!}{2!} - \frac{n!}{3!} + \cdots + (-1)^n \frac{n!}{n!} = \text{round}(n!/e).$$

(b) To construct the desired permutation, select those 4 books from 10 that remain in their original places, and then rearrange the remaining books, so that no book remains in its original place. Using (a), the required quantity is equal to

$$\binom{10}{4}\left(6! - \frac{6!}{1!} + \frac{6!}{2!} - \frac{6!}{3!} + \cdots + \frac{6!}{6!}\right)$$

$$> \frac{10 \cdot 9 \cdot 8 \cdot 7}{24}\left(\frac{6!}{2} - \frac{6!}{6}\right) = 210 \cdot 240 = 7 \cdot 7200 > 50\,000.$$

1.3.7. (a) Let U be the set of all tourist arrangements, and let A_j be the set of tourist arrangements for which the jth house is empty.

The number of arrangements for which all houses with numbers from the set S are empty is equal to $\left|\bigcap_{j\in S} A_j\right| = (5 - |S|)^{20}$. Inclusion-exclusion yields

$$|U - (A_1 \cup \cdots \cup A_5)| = \sum_{k=0}^{5}(-1)^k \binom{n}{k}(5-k)^{20}$$

$$= 5^{20} - \binom{5}{1}4^{20} + \binom{5}{2}3^{20} - \binom{5}{3}2^{20} + \binom{5}{4}1^{20}.$$

(b) Let $U = \mathbb{Z}_n^{\mathbb{Z}_k}$ be the set of all mappings from $\mathbb{Z}_k$ to $\mathbb{Z}_n$. For each $j = 1, \ldots, n$, let $A_j = (\mathbb{Z}_n - \{j\})^{\mathbb{Z}_k}$ be the set of all mappings $\mathbb{Z}_k$ to $\mathbb{Z}_n$, whose image does not contain j. Then the set of surjections from $\mathbb{Z}_k$ to $\mathbb{Z}_n$ is the set $U - (A_1 \cup A_2 \cup \cdots \cup A_n)$. Observe that $\left|\bigcap_{j\in S} A_j\right| = (n - |S|)^k$. Inclusion-exclusion yields

$$|U - (A_1 \cup A_2 \cup \cdots \cup A_n)|$$

$$= n^k - \binom{n}{1}(n-1)^k + \binom{n}{2}(n-2)^k - \cdots + \binom{n}{n} \cdot 1^k.$$

1.3.8. (a) *Answer:* $\frac{n}{n-k}\binom{n-k}{k}$.

(b) Use part (a).

1.3.9. *Answer:* 333.

First suggestion. Inclusion-exclusion yields $400 - 3 \cdot 20 + 3 \cdot 1 - 10 = 333$.

Second suggestion (from site http://problems.ru). One horizontal layer G and two vertical layers pass through a given unit cube K. The sum of all numbers in the 361 vertical columns that are not in the other two vertical layers is 361. Then we must subtract the sum S of labels of the 361 cubes lying in the intersection of these columns with G. These cubes are completely covered by 19 columns lying in G. The sum of all the labels in these columns (it is equal to 19) exceeds S by the sum of the 19 labels lying in the column containing K that is perpendicular to them. The last amount is obviously $1 - 10 = -9$. Hence $S = 19 - (-9) = 28$. Our final sum is thus $361 - 28 = 333$.

Chapter 2

Finite sets

This topic can be accessible for students in grades 6–7, but we suggest that instead of this chapter, you consult the relevant sections in the book [**FGI**]. General methodological guidelines are also given there.

1. The pigeonhole principle (1)
By A. Ya. Kanel-Belov

Part 1

The pigeonhole principle is a very simple statement. Here are three formulations.

2.1.1. (a) **The pigeonhole principle.** If more than n rabbits occupy n cages, then there is a cage containing at least two rabbits.

(b) **The generalized pigeonhole principle.** If N rabbits occupy n cages, then at least one cage contains at least N/n rabbits.

(c) **The continuous pigeonhole principle.** If the sum of n real numbers is N, then at least one number is greater than or equal to N/n and at least one number is less than or equal to N/n.

Surprisingly, these simple ideas turns out to be crucial in many difficult problems.

Let's start with easy problems.

2.1.2. A 10-question quiz was taken by 27 students. Prove that there are 3 students who solved the same number of problems and that there are 7 students who got the same score. (Quiz grade can be $5, 4, 3, 2$.)

2.1.3. On Earth, there are more than 7 billion people, of whom more than 99% are less than 100 years old. Prove that there are 2 people on Earth born within an interval of less than one second.

2.1.4. Prove that among any 11 numbers you can find 2 whose difference is divisible by 10.

2.1.5. Prove that among any 5 people there are 2 with the same number of acquaintances among these 5 people.

2.1.6. Prove that among any 5 numbers there are several numbers whose sum is divisible by 5.

2.1.7. On the plane, there are selected 5 distinct points with integer coordinates. Prove that the midpoint of one of the segments connecting these points also has integer coordinates.

2.1.8. Twenty-one boys together gathered 200 nuts. Prove that there are two boys who gathered the same number of nuts.

2.1.9. Prove that you can write a multiple of 1991 using only the digit 1.

2.1.10. In each cell of a 5×5 board sits a bug. Suddenly, each bug jumps to a horizontally or vertically adjacent cell. Prove that at least two bugs will land in the same cell.

2.1.11. Prove that from any $k + 2$ natural numbers, where $k \geq 1$, you can choose two numbers whose sum or difference is divisible by
    (a) 2;   (b) $2k + 1$.

2.1.12. On the plane, 12 lines are drawn. Prove that there are two of them that form an angle of at most $\pi/12$.

2.1.13. Six asterisks are placed in the cells of the 4×4 table. Prove that it is possible to cross out two rows and two columns from the table so that there will be no asterisks in the remaining cells.

Part 2 (2)

Now we turn to more complex problems, where the pigeonhole principle is combined with other ideas.

2.1.14. In the game of *Battleship*, what is the fewest number of shots that guarantees you will hit a four-cell ship?

2.1.15. Each of nine lines divides a square into two quadrilaterals whose areas are in the ratio 2 : 3. Prove that at least three of these lines are concurrent.

2.1.16. Is it possible to label the vertices of a regular 45-gon with the numbers $0, 1, \ldots, 9$ so that for any pair of different numbers, there is a side whose endpoints are labeled with these numbers?

2.1.17. A committee of 60 people held 40 meetings, and each meeting was attended by exactly 10 committee members. Prove that two members were at the same meeting at least twice.

2.1.18. Prove that among any $2k + 1$ different integers of absolute value at most $2k - 1$, you can find three numbers whose sum is zero.

Now for harder problems.

2.1.19.* Consider an infinite sequence $\{a_n\}$ of natural numbers greater than 1, where no number appears in the sequence $\{a_n\}$ twice. Prove that there are infinitely many n such that $a_n > n$.

2.1.20.* On a 1000×1000 chessboard, place a black king and 499 white rooks. Prove that for any initial arrangement of the pieces, the king will get in check by a white rook, no matter how white plays.

2.1.21.* On a table lie 50 properly running watches. Prove that at some time, the sum of the distances from the center of the table to the ends of the minute hands will exceed the sum of the distances from the center of the table to the centers of the clocks.

The so-called pigeonhole principle for lengths and areas is described in Sections 4 and 5 in Chapter 7.

Suggestions, solutions, and answers

2.1.2. Each student solved between 0 and 10 problems, inclusive; that is, there are 11 total possible values. There are 27 students; therefore, by the generalized pigeonhole principle there are at least $\frac{27}{11}$ students with the same number of problems solved. Since $\frac{27}{11} = 2\frac{5}{11} > 2$, there are more than two (that is, at least three) students who have solved the same number of problems. Similarly, the number of students who received the same grade (out of four possible grades) is at least $\frac{27}{4} = 6\frac{1}{4} > 6$, that is, at least 7.

2.1.5. Note that the number of acquaintances of each person (among this group of 5) can be any number from 0 to 4 inclusive (only 5 options). Observe that a person with 4 acquaintances and a person with 0 acquaintances cannot both be in the company at the same time since the former would know everyone, and the latter would not know anyone, which is impossible. This means that at least one of these two options (0 acquaintances and 4 acquaintances) does not occur, and no more than 4 options remain for the group. Since there are 5 people in the group, the pigeonhole principle implies that two people have the same number of acquaintances.

2.1.9. See article [**Kor**].

2.1.13. First, we show that it is possible to cross out two columns that contain at least 4 asterisks: by the pigeonhole principle, in a column with the maximum number of asterisks there are at least $\frac{6}{4} = 1\frac{1}{2} > 1$, that is, at least 2 asterisks. If there are 3 or 4, then we take as the second column an arbitrary one containing at least one asterisk. In two selected columns there will be at least 4 asterisks, which we need. If the first column contains exactly 2 asterisks, then the remaining 3 columns contain 4 asterisks; the pigeonhole principle implies that among them there is a column containing at least 2 asterisks. Take this as the second column and again we get two columns with 4 asterisks.

In the remaining two columns, no more than two asterisks are left. They occupy no more than two lines. We delete these two lines and get the desired result.

2. The extremal principle (2)
By A. Ya. Kanel-Belov

The key to solving many problems is to consider "extremal" values of some of the entities in the problem. This may be the largest of the numbers in the problem, the segment with the longest length, the circle with the smallest radius, and so on.

2.2.1. Place numbers at the vertices of a regular 100-gon, each of which equals the arithmetic mean of their two neighbors. Prove that all the numbers are equal.

2.2.2. Numbers are placed in each square of a chessboard so that each number is equal to the arithmetic mean of the numbers in neighboring squares (sharing sides). Prove that all the numbers are equal.

2.2.3. The vertices of the cube are numbered $1, 2, \ldots, 8$. Prove that there is an edge whose endpoint labels differ by at least three.

2.2.4. Prove that any 10 points in the plane are endpoints of some 5 non-intersecting segments.

2.2.5. Prove that it is not possible to arrange 100 distinct points in space so that each point lies at the midpoint of the line segment joining two other points.

2.2.6. Perpendiculars are dropped from a point inside a convex polygon to each of its sides or to the extension of a side. Prove that at least one perpendicular must land on a side, not on the extension of it.

2.2.7. On each of the planets of a certain planetary system sits an astronomer who observes the nearest planet. The distances between planets are pairwise distinct. Prove that if the number of planets is odd, then some planet will not be observed.

2.2.8. Several identical coins lie on a table. Prove that at least one coin is tangent to at most three other coins.

2.2.9. In the kingdom of Far Far Away, every two cities are connected by a one-way road. Prove that there is a city from which you can reach any other city by traveling on no more than two roads.

2.2.10. Distinct integers are placed in the cells of as 8×8 chessboard. Prove that there are two neighboring (having a common side) cells whose numbers differ by at least 5.

2.2.11. A number is written in each cell in an infinite grid. Prove that there is a number that is less than or equal to at least four of eight numbers in the neighboring cells.

Suggestions, solutions, and answers

2.2.1. Consider the maximum of the numbers. It is equal to half the sum of its neighbors and is no less than each of them. Consequently, each of the neighbors of this number is equal to this maximum number. Similarly, their neighbors are equal to the maximum number, etc. Repeating this argument 50 more times implies that all the numbers are equal.

2.2.4. Consider all the ways to connect these points with 5 segments, and focus on the configuration for which the total length of the segments is minimal. Another way to solve this problem is discussed in Section 5 in Chapter 4; in particular, see the solution to problem 4.5.8.

2.2.5. This problem (in an equivalent formulation) is discussed in detail in [**Ros**].

2.2.7. We use induction on the number of planets. For three planets the statement is obvious. If every planet is observed by someone, then no more than one astronomer observes the planet. Consider the two closest planets. Their astronomers look at each other, and no one else watches them. Now apply the induction hypothesis to the remaining planets.

3. Periodicity I (2)
By A. Ya. Kanel-Belov

Problems in this section follow [**K-BS**].

2.3.1. Find the last digit of 2^{1000}.

An infinite sequence of numbers $a_1, a_2, \ldots$ is called *cyclic* (or *periodic*) if, starting at some term, the sequence then is just an infinite repeat of one finite sequence. For example, $1, 2, 3, 4, 1, 2, 3, 4, \ldots$ is a periodic sequence.

The repeated finite sequence is called the *period* of the sequence. In the example above the period is clearly not unique: both $1, 2, 3, 4$ and $1, 2, 3, 4, 1, 2, 3, 4$ (and also infinitely many others) are periods.

The repeating part of a periodic sequence may not start at the first term; there may be a number of terms before the repeats. The *preperiod* is the shortest initial term of the sequence after which it repeats. For example, $2, 2, 1, 2, 3, 1, 2, 3, 1, 2, 3 \ldots$ is a sequence with a preperiod of $2, 2$ and period $1, 2, 3$.

2.3.2. Prove that the remainders of the powers of 7 upon division by 1990 form a periodic sequence and that the length of its smallest period is at most 1990.

2.3.3. The inhabitants of the Lawaiian Islands pride themselves on having, from times immemorial, a presidential form of government. Every 4 years, either a Republican or a Democrat is elected president. Local sociologists have discovered a strict law that predicts the partisanship of the next president. Although this law is classified, information has leaked into the press that the party membership of the next president is fully determined by the partisanship of the previous ten. The last three presidents were Republican. Prove that there will be an infinite number of repeats of the rule of three Republicans in a row.

2.3.4.* The Fibonacci sequence $\{f_n\}$ is defined by the conditions $f_1 = f_2 = 1$ and $f_{n+1} = f_n + f_{n-1}$ for $n \geq 2$. Prove that there is a member of the Fibonacci sequence divisible by 1990.

2.3.5.* *Recurrent sequences.* A sequence $u_1, u_2, \ldots$ is called recurrent if for some $k, a_1, \ldots, a_k$ and all $n > k$, the equality $u_n = a_1 \cdot u_{n-1} + a_2 \cdot u_{n-2} + \cdots + a_k \cdot u_{n-k}$ holds. Suppose all a_i and u_n are integers.

(a) Prove that for each integer l, the remainders of u_n when divided by l are periodic.

(b) Estimate the length of this period.

(c) Give an example for which there is a nonempty preperiod.

(d) Find a criterion for the presence of a nonempty preperiod.

The idea of periodicity may appear in different guises.

2.3.6. A combination of face rotations scrambled a Rubik's Cube from its initial correct position. Prove that this initial position can be restored by repeating this combination several more times.

2.3.7. A mathematics olympiad had 20 problems, and after the competition, 20 students gathered to discuss their solutions. Each of them solved two problems, and it turned out that each problem was solved by two students among these 20 students. Prove that you can organize the discussion so that each student explains the solution of one of the problems he solved and all problems are discussed.

2.3.8. Each term of a sequence, starting with the second term, is equal to the sum of the squares of the digits of the previous term. Prove that this sequence is periodic.

2.3.9. An English text is printed on an infinite tape (infinite in both directions). It turns out that the number of distinct 15-character strings is equal to the number of distinct 16-character strings. Prove that the text must be *periodic*; for example: ... Mom washed the window frame Mom washed the window frame....

Let us turn to more complex problems requiring research.

2.3.10. There are an unlimited number of black and white cubes, from which we wish to build a continuous tower in the shape of a parallelepiped so that each black cube shares a face with an even number of white cubes, and each white cube shares a face with an odd number of black cubes. Is it true that any arrangement of cubes on the "first floor" (i.e., the lowest layer of the tower, with a height of one cube) can be extended to such a tower of a finite height?

2.3.11.* A strip of grid paper, endless in both directions, consists of black and white cells. Every second each cell with an even number of black neighbors turns white, and each cell having an odd number of black neighbors turns black. Prove that:

(a) If after 2^n seconds the original coloring is repeated, then it is periodic with period $3 \cdot 2^n$.

(b) The original coloring is periodically repeated if and only if it is itself periodic (i.e., periodicity in time is equivalent to periodicity in space).

2.3.12.* Starting with any number, we append digits on the right, one at a time. We can append any digit except for 9. Prove that sooner or later you get a composite number.

2.3.13. One is given a rectangle with the side ratio equal to $\sqrt{7}$. From it we cut the square of side equal to the shorter side of the rectangle. With the remaining rectangle we perform the same procedure. Prove that the sequence of side ratios of the constructed rectangles is periodic.

Suggestions, solutions, and answers

Problems 2.3.1–2.3.4 and 2.3.6 are considered in details in article [**K-BS**].

2.3.1. Consider the sequence of the last digits of the powers of two: $1, 2, 4, 8, 6, 2, 4, 8, 6, \ldots$. Note that each term in the sequence completely determines the next term. Therefore, after any 2 there will be a 4 and so on, so $2, 4, 8, 6$ is the period. Then it is easy to determine that 2^{1000} ends with 6.

2.3.4. Write down the sequence of remainders of the Fibonacci numbers upon division by m: $r_1 = 1, r_2 = 1, \ldots$. There is a finite set of ordered pairs of remainders, so there are numbers $i \neq j$ such that $r_i = r_j$ and $r_{i+1} = r_{j+1}$. The remainders r_i and r_{i+1} uniquely determine subsequent and previous remainders; therefore the sequence of the remainders is periodic with period $l := j - i$ (without a preperiod). The periodicity implies that $r_{l+1} = r_{l+2} = 1$, so $r_l = 0$.

4. Periodicity II (2) *By P. A. Kozhevnikov*

2.4.1. Solve the system of equations $x_1 + \sqrt{x_2} = x_2 + \sqrt{x_3} = \cdots = x_n + \sqrt{x_1} = 2$ for $n = 9$.

2.4.2. In a given convex polygon, all angles are equal. The polygon has an interior point from which all sides subtend at equal angles. Prove that this polygon is regular. (*K. Kokhas*, The problems from "Kvant", 1988, Nos. 11–12.)

2.4.3. Twenty teams participate in a round robin football championship. Prove that after two rounds, you can choose ten teams, among which no two have played each other. (*S. Genkin*, Tournament of Towns, 1986.)

2.4.4. An infinite sequence of digits $9, 6, 2, 4, \ldots$ is constructed by the following rule: each digit is equal to the last digit of the sum of the previous four digits. Will the sequence $2, 0, 0, 7$ ever appear in this sequence?

2.4.5. A five-pointed star is obtained by intersecting five straight lines (no two lines are parallel, and no three lines coincide). The boundary of the star consists of ten segments. Color each boundary segment red or blue, alternating as you go around the star. Could it happen that each red segment is longer than any blue segment?

2.4.6. A circular necklace is made of n beads, colored red and blue. For a fixed number k, we call the necklace *good* if between any two red beads there are never exactly $k - 1$ beads. For the following cases, find the largest number of red beads that a good necklace can have: (a) $n = 15$, $k = 4$; (b) $n = 15$, $k = 6$; (c) any n and $k < n$.

2.4.7. In a certain city, each family lives in a separate apartment, but they are permitted to swap apartments with another family, provided that they do not do any other types of switching and they cannot do more than one swap per day. Prove that any permutation of apartments among any number of families in this city can be completed in two days. (*A. Schnirelman and N. Konstantinov* Tournament of Towns, 1987.)

2.4.8. N point size balls move around the circle. Each ball has speed v (clockwise or counterclockwise). If two balls collide, then they scatter in opposite directions so that their speeds remain equal to v. Prove that eventually all the balls will return to their initial position at the same time.

Suggestions, solutions, and answers

2.4.1. Note that $x_1 = x_2 = \cdots = x_n = 1$ is a solution. Suppose that $x_1 > 1$. From the equalities we successively obtain that $x_2 < 1$, $x_3 > 1$, ..., $x_9 > 1$, $x_1 < 1$, which is a contradiction. Likewise, we arrive at a contradiction if we assume that $x_1 < 1$.

Comment. In fact, $x_1 = x_2 = \cdots = x_n = 1$ is the only solution regardless of the parity of n. The equations $x_i + \sqrt{x_{i+1}} = 2$ $(i = 1, 2, \ldots, n-1)$ imply that for $x_1 \neq 1$ the distance from x_k to 1 increases with increasing k.

2.4.2. Let $A_1 A_2 \ldots A_n$ be a polygon, and let O be a point inside it. From

$$\angle A_1 A_2 A_3 = \angle A_2 A_3 A_4 = \cdots = \angle A_n A_1 A_2 = \frac{\pi(n-2)}{n},$$

$$\angle A_1 O A_2 = \angle A_2 O A_3 = \cdots = \angle A_n O A_1 = \frac{2\pi}{n}$$

it follows that $\angle O A_1 A_2 = \angle O A_2 A_3 = \cdots = \angle O A_n A_1$. Triangles $O A_1 A_2$, $O A_2 A_3$, ..., $O A_n A_1$ are similar; therefore $O A_1 / O A_2 = k$, $O A_2 / O A_3 = k$, ..., $O A_n / O A_1 = k$, where k is the coefficient of similarity. Multiplying the equalities we get $k = 1$, and, therefore, $O A_1 = O A_2 = \cdots = O A_n$.

2.4.3. Construct a graph of the played games. Two edges come out of each vertex; therefore, the graph is a union of disjoint cycles (round dances!). These are cycles of even length (a cycle of odd length cannot be played in two rounds). From each cycle, just select half of the teams that did not play with each other.

2.4.4. *Answer*: the sequence $2, 0, 0, 7$ will occur.

Suggestion. Each 4-digit block determines both the next *and* the previous digit in this sequence. Since there are only finitely many ordered 4-digit blocks, eventually two identical blocks will appear. This implies that the sequence is periodic and has no preperiod. Thus the block $9, 6, 2, 4$ will repeat, not just appearing at the start. It is easy to see that the digits preceding $9, 6, 2, 4$ must be $2, 0, 0, 7$.

2.4.5. *Answer*: cannot.

Hint. For five triangles containing a red side and a blue side, apply the theorem that the angle opposite the larger side of a triangle has the larger angle.

2.4.6. (c)*Answer*: $d\left\lfloor \frac{n}{2d} \right\rfloor$, where $d = \text{LCM}(n, k)$.

Suggestion. Connect beads with an edge if there are $k-1$ beads between them. This is the "prohibition graph": two red beads cannot be connected by an edge. Let $d = \text{GCD}(n, k)$. Then the graph decomposes into d cycles of length n/d. In each cycle, there can be no more than half of the red bead, and exactly $\left\lfloor \frac{n}{2d} \right\rfloor$ red beads can be painted.

2.4.7. Any permutation is a union of *disjoint* cycles. It remains to be show that a cycle can be carried out in two days: any rotation can be represented as a sequential application of two axial symmetries.

2.4.8. We show that after each time interval d/v, where d is the length of the circumference, at the initial position of each ball A there will be found some ball B (not necessarily $A = B$).

Indeed, if we assume that the balls are indistinguishable, then we can assume that no collisions occur (that is, at the time of the collision, the balls "pass through each other"), and thus, in time d/v the balls will resume their initial positions, and in each position the direction of the velocity of the ball coincides with the initial direction of the velocity of the ball in this position.

Now recall that the balls are different. Then in time d/v there is a permutation of the balls. Some power of this permutation is the identity permutation.

Comment. The structure of this permutation can be determined knowing the number of balls that move in clockwise and counterclockwise directions (see Problem M2291 from [**PROB2**]).

5. Finite and countable sets (2)
By P. A. Kozhevnikov

2.5.1. The set of natural numbers is partitioned into infinite arithmetic progressions with differences $d_1, d_2, \ldots$.

(a) Prove that if the number of progressions is finite, then $\frac{1}{d_1} + \frac{1}{d_2} + \cdots = 1$.

(b) Is (a) true if the number of progressions is infinite? (*A. Tolpygo*, Tournament of Towns, 1989.)

2.5.2. The entire plane is illuminated by spotlights, each illuminating an angle with vertex at the spotlight.

(a) Prove that if the number of spotlights is finite, then the sum of the angles is at least 360°.

(b) Is (a) true if the number of angles is infinite (countable)?

2.5.3. Is it possible to place all natural numbers in each cell of an infinite 2-dimensional grid so that each number occurs exactly once and any two numbers in a row or column are relatively prime?

2.5.4. The set of natural numbers is partitioned into two sets A and B. It is known that A does not contain a three-term arithmetic progression. Could it be that B does not contain an infinite arithmetic progression? (*A. Skopenkov*, Problems of *Kvant*, 1990, No. 8.)

2.5.5. Is there a subset M of the set of natural numbers such that each natural number in M is uniquely represented as a difference of two numbers from M?

2.5.6. Is it possible to color integer points of the plane in 2007 colors so that all colors are present on each line (that passes through at least two, and thus infinitely many integer points) and the coloring on each line is periodic, yet the coloring on the plane was not periodic? (The coloring of the plane is periodic if there is a nonzero integer vector, such that when you translate by this vector, the coloring of all points is invariant.)

2.5.7. Does there exist a function $f\colon \mathbb{Q} \times \mathbb{Q} \to \mathbb{Q}$, which is *not* a polynomial in two variables, yet for any $a \in \mathbb{Q}$, both functions $f(a,x)$ and $f(x,a)$ are polynomials?

2.5.8. Is it possible to arrange queens on an infinite chessboard so that each of them threatens exactly 7 others?

2.5.9. Does there exist a sequence $a_1, a_2, \ldots$ of natural numbers such that every natural number appears in it exactly once and for each n, the sum $a_1 + \cdots + a_n$ is divisible by n?

Suggestions, solutions, and answers

2.5.1. (a) Let $a_1, a_2, \ldots, a_n$ be the first members of the progressions. Take N consecutive natural numbers $A + 1, A + 2, \ldots, A + N$, where $A > \max\{a_1, a_2, \ldots, a_n\}$ and $N = d_1 d_2 \ldots d_n$. It is easy to see that among the selected N numbers, exactly $\frac{N}{d_i}$ numbers (exactly one for each d_i consecutive numbers) lie in the ith progression. It follows that $N = \frac{N}{d_1} + \frac{N}{d_2} + \cdots + \frac{N}{d_n}$. Dividing both sides by N, we get the desired result.

(b) *Answer*: no.

Suggestion. We will construct a partition of the natural numbers that satisfies $\frac{1}{d_1} + \frac{1}{d_2} + \cdots = \frac{1}{2^k}$, where k is a fixed natural number. To do this, we construct progressions with differences $d_i = 2^{i+k}$.

Here is a way to construct the first term a_i of the ith progression. Let $a_1 = 1$. If $a_1, \ldots, a_n$ are already selected, then let a_{n+1} be the smallest natural number not included in any of the progressions (it can be shown that such a number exists even among the first d_n numbers). Since d_{n+1} is divisible by $d_1, d_2, \ldots, d_n$, the $(n+1)$st progression is disjoint from the previously defined progressions (prove it!).

2.5.2. (a) Using parallel translations, move the spotlight so that they all have the common vertex O. Let the sum of the angles be less than $360°$. Then there is a ray starting at O that is not illuminated by the translated spotlights. It is easy to prove that before the translation each spotlight illuminated a limited part on this ray (a segment or an empty set), so part of the ray was unlit; a contradiction.

(b) *Answer*: not true.

Suggestion. Let us describe the covering with angles of size $\alpha_i = \left(\frac{1}{2^{i+k}}\right)^\circ$ so that

$$\sum_{i=1}^{\infty} \alpha_i = \left(\frac{1}{2^k}\right)^\circ.$$

Consider circles with radii $1, 2, \ldots$ centered at the origin. Clearly, if each of these circles is covered by some spotlight, then the whole plane will be covered. Now place the ith spotlight far enough from the origin to cover the circle with radius i.

For problems 2.5.3–2.5.9, the answer is always positive. As in problems 2.5.1(b), 2.5.2(b), the object whose existence is to be proved can be "constructed" in a countable number of steps. Suppose that after the nth step, we have constructed the set K_n. At the $(n+1)$st step, the set K_n "is extended" to the set $K_{n+1} \supset K_n$. The desired construction is the union $K = \bigcup_{i=1}^{\infty} K_i$.

Below we use the fact that for countable sets A_i the set $A_1 \times A_2 \times \cdots \times A_k$, i.e., the set of ordered k-tuples $\{(a_1, a_2, \ldots, a_k) : a_i \in A_i\}$, is countable.

2.5.3. Label the cells arbitrarily. Arrange the numbers 1, 2, 3, ... in order, guided by the following rule. In the first step, put the number 1 in the cell labeled 1. Suppose that after n steps, the numbers $1, 2, \ldots, n$, are already placed; let k be the minimum label of the cell in which there is no number yet. Put the number $n+1$ in the cell with label k if $n+1$ is prime. Otherwise, we put $n+1$ in any cell with no other numbers in its row and column. Eventually, the cell with the label k will be occupied, since there are infinitely many primes.

2.5.4. We will arbitrarily enumerate all infinite arithmetic progressions consisting of natural numbers (there are countably many, since each progression is determined by its first term and difference). Construct $A = \{a_1, a_2, \ldots\}$, where the numbers $a_1, a_2, \ldots$ are defined as follows. Let a_1 be any number from the 1st progression. If a_n is already defined, then select a_{n+1} from the $(n+1)$st progression such that $a_{n+1} > 2a_n$ holds. It is easy to see that the sets A and $B := \mathbb{N} - A$ satisfy the condition.

2.5.5. We describe the construction of the desired sequence $M = \{a_1, a_2, \ldots\}$. For each natural number k, define an initial piece A_k of the sequence $a_1 < a_2 < \cdots < a_m$ (where $m = m(k)$) so that differences of the form $a_j - a_i$ for $1 \le i < j \le m$ are pairwise distinct and the set R_k of these differences contains the numbers $1, 2, \ldots, k$. If R_k also contains the number $k+1$, then we do nothing; i.e., put $A_{k+1} = A_k$. Otherwise, we add two numbers a_{m+1} and $a_{m+2} = a_{m+1} + k + 1$ such that $a_{m+1} > 10a_m$. Then all differences of the form $a_{m+1} - a_i$, where $i = 1, 2, \ldots, m$, are pairwise distinct, greater than any number from R_k and greater than $k+1$. The same is true for differences of the form $a_{m+2} - a_j$, where $j = 1, 2, \ldots, m$. The equality of the differences $a_{m+1} - a_i = a_{m+2} - a_j$ for $i, j \in \{1, 2, \ldots, m\}$ implies that $a_j - a_i = k+1$; a contradiction. Thus, we have extended A_k to A_{k+1}, fulfilling the necessary conditions.

2.5.6. We enumerate the lines containing infinitely many integer points and enumerate the nonzero integer vectors (both sets are countable). In the first step, we color the first line periodically using 2007 colors. On the ith $(i \ge 2)$ step we periodically color the ith line (this is possible, since only finitely many points on it have been painted), and then we find two points in the plane that have not yet been painted, separated from each other by the $(i-1)$st vector, and color them in different colors. It follows from the definition of the procedure that the color of each line is periodic, but the color of the plane cannot be periodic, since no translation by an integer vector will leave all the colors invariant.

2.5.7. We describe the construction of the desired function f. Enumerate the rational numbers $q_1, q_2, \ldots$. In the first step, set $f(q_1, y) = f(x, q_1) = 0$ for all $x, y \in \mathbb{Q}$. At the ith $(i \ge 2)$ step we will set the values of $f(x, y)$ on the lines $x = q_i$ and $y = q_i$. Before the ith steps, the values of the function f are already determined only in the finite number of points on the line $x = q_i$ (namely, at points $(q_i, q_1), (q_i, q_2), \ldots, (q_i, q_{i-1})$). We choose a polynomial $P_i(y)$ of degree greater than i such that $P_i(q_k) = f(q_i, q_k)$ for $k = 1, 2, \ldots, i-1$ (such a polynomial exists; this can be proved, for example, by using *Lagrange interpolation*), and we set $f(q_i, y) = P_i(y)$. Further, let $f(x, q_i) = h_i(x)$, where $h_i(x)$ is a polynomial of a degree greater than i such that $h_i(q_k) = f(q_k, q_i)$ for $k = 1, 2, \ldots, i$. This completes the ith step of construction.

If the function $f(x, y)$ we constructed was a polynomial in two variables, then the degrees of the polynomials $f(x, q_i)$ $(i = 1, 2, \ldots)$ would be bounded, a contradiction.

Comments about the solutions of problems 2.5.1 and 2.5.2

2.5.1. Let M be a subset of the natural numbers. Consider the sequence $q_n = \frac{k(n)}{n}$, where $k(n)$ is the number of members of the set M among the numbers $1, 2, \ldots, n$. If the limit $Q = \lim_{n \to \infty} q_n$ exists, then we call Q a *measure* or *density* of the subset M, and we denote it by $\mu(M)$.

(Of course, this concept does not coincide with the classical definition of measure spaces. For example, with our definition the union of two *measurable* sets, i.e., sets for which a measure is defined, is not necessarily measurable.)

In a sense, the measure $\mu(M)$ represents the probability that a randomly selected natural number will be a number from the set M.

It is easy to verify the following properties of this measure:

- $\mu(\mathbb{N}) = 1$;
- the measure of a finite set is 0;
- the measure of an infinite arithmetic progression with difference d equals $1/d$.

Consider the following problems:

(**c**) show that there are nonmeasurable subsets;

(**d**) find the measure of the set of exact squares;

(**e**) find the measure of the set of primes.

It is easy to prove that μ is additive, i.e., for measurable disjoint subsets A and B, the set $A \cup B$ is measurable, and $\mu(A \cup B) = \mu(A) + \mu(B)$. The additivity of the measure μ implies equality in problem 2.5.1(a).

However, the measure μ is not σ-*additive*; i.e., for pairwise disjoint measurable subsets of A_1, A_2, $\ldots$, the equality $\mu\left(\bigcup_{i=1}^{\infty} A_i\right) = \sum_{i=1}^{\infty} \mu(A_i)$ is not always satisfied (and in general, $\bigcup_{i=1}^{\infty} A_i$ may not be measurable). For example, $\mathbb{N}$ is a union of sets each consisting of one number, but $1 = \mu(\mathbb{N}) \neq \sum_{i=1}^{\infty} \mu(\{i\}) = 0$. An example from problem 2.5.1(b) also proves the absence of σ-additivity of the measure μ.

2.5.2. Analogously to the previous problem, we can try to define the measure of some subsets M of the plane. Let $q_R = \frac{S_R}{\pi R^2}$, where S_R is the intersection area of $M \cap C_R$ with the circle C_R of radius R with the center at the origin (definition of the area discussed in Section 5 of Chapter 7). If the limit $Q = \lim_{R \to +\infty} q_R$ exists, then we call Q the *measure* of the subset M, denoted by $\mu(M)$.

In a sense, the measure $\mu(M)$ is the probability that a randomly chosen point on the plane will be a point from the set M.

It is easy to verify the following properties of a measure:

- the measure of the entire plane is 1;
- the measure of a bounded subset of a plane is 0;
- the measure of the interior of the angle of α° is $\alpha/360$.

Consider the following problems:

(**c**) Show that there are nonmeasurable subsets of the plane.

(**d**) Will the concept of measurability and measure change if we change the location of the origin?

(**e**) Will the notion of measurability and measure change if C_R are not circles, but, say, the squares $|x| + |y| \leq R$?

As in problem 2.5.1, measure μ turns out to be additive, but not σ-additive. Ponder another solution to the problem 2.5.2(a).

Chapter 3

Graphs
By D. A. Permyakov
and A. B. Skopenkov

This chapter contains materials for an introduction to graphs. We recommend continuing the study of graphs in Chapters 2, 3, and 4 of [**GDI2**].

1. Graphs (2)

Solving problems from this and the next section does not require any prior knowledge and is suitable for an introduction to graphs.

3.1.1. Is it possible, after making several moves with the chess knights from the starting position shown in Fig. 1 on the left, to arrange them as shown in Fig. 1 on the right?

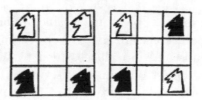

FIGURE 1

3.1.2. (a) Draw how to connect 12 computers by cables in a network so that the following conditions are met:

- each cable connects two computers;
- exactly 3 cables are connected to each computer;
- any computer can send a message via cable to any other computer so that it passes through at most 2 other computers along the way.

(b) The same, only instead of computers, processors on one side of the printed circuit board are considered; it is additionally required that the cables are located on the same side of the circuit board and do not cross each other.

3.1.3. (a) Some pairs of (a finite number of) cities are connected by (non-stop) flights. Are there necessarily two cities from which you can fly nonstop to the same number of cities?

(b) A foreign intelligence agent reported that each of the 15 former republics of the USSR concluded an agreement with exactly 3 others. Can you trust him?

(c) In a kingdom there is the capital, the city of Dalniy, and several other cities. Some pairs of cities are connected by roads (roads may intersect). There are 21 roads that leave the capital, one road that leaves Dalniy, and 20 roads that leave each of the remaining cities. On each road, there is a bus line. Is it possible to ride a bus from the capital to Dalniy (possibly with transfers)?

3.1.4. A convex polygon (different from a triangle) is divided into triangles by several diagonals that do not intersect anywhere except at vertices. A triangle is called *good* if two of its sides are sides of the original polygon. Is it true that there always exist at least two good triangles?

3.1.5. Is it true that in any company of six people there are either three pairwise acquaintances or three pairwise unfamiliar people?

The following problem is interesting only if the solution contains rigorous reasoning.

3.1.6. In a group of several people, some people are familiar with each other (i.e., they are friends), and some are not. Every evening one of them arranges dinner for all his friends and introduces them to each other. After each person has arranged at least one dinner, it turns out that there are two people who still are not friends. Prove that at the next dinner they will not be able to meet either.

3.1.7. (a) In the Duma there are 450 deputies, each of whom threatens exactly one other deputy. Prove that you can choose 150 deputies so that among them no one threatens any other.

(b) In an infinite group of gangsters, each one hunts after exactly one other. Prove that there is an infinite subset of gangsters in which no one is hunting for any other.

(c) In elections to the City Duma, each voter, if he comes to the polls, votes for himself (if he is a candidate) and for all those candidates who are his friends (the list of candidates and their friends is fixed). The forecast of the political service of the mayor's office is considered *good* if it correctly predicts the number of votes cast for at least one of the candidates, and it is *not good* otherwise. Prove that with any forecast, voters can turn up in such a way that this forecast will turn out to be *not good*.

3.1.8. A tourist who arrived in Moscow by train wandered around the city all day. Having dined in a cafe in one of the street intersections she decided to return to the train station along the streets which she had walked an odd number of times before. Prove that she can always do this.

3.1.9. In some community, any two acquaintances do not have common acquaintances, but any two strangers have exactly two common acquaintances. Prove that in this community everyone has the same number of acquaintances.

Suggestions, solutions, and answers

Some solutions in this section were written by S. Erlykov and edited by A. Gavrilyuk. It is assumed that the reader has already become familiar with the definitions introduced at the beginning of Section 3 of this chapter, "Paths in graphs".

3.1.1. *Answer*: no. *Solution*. See [**FGI**, ch. 6, § 1, problem 2].

3.1.2. (a), (b) See Fig. 2. This and other similar problems are discussed in [**CF**].

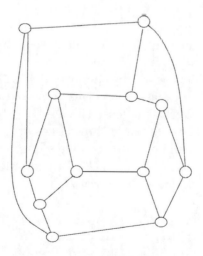

FIGURE 2. Graph in problem 3.1.2.

3.1.3. (a) *Answer*: yes, there certainly are.

Solution. We associate with each city a vertex of the graph, and with each flight an edge connecting the corresponding vertices (cities). We get a graph with n vertices. (However, the following solution can also be reformulated starting with flights.)

Assume first that each vertex is an endpoint of some edge. Then the maximum possible number of edges emerging from one vertex is $n - 1$, and the minimum possible is 1, so that there are $n - 1$ options for the degree of an arbitrary vertex. Since there are n vertices in the graph, there are two vertices for which the degrees are equal, that is, from which the same number of edges comes out.

If there are isolated vertices (with degree 0), then we exclude them from consideration. The problem reduces to the previous case.

Comment. In essence, this is problem 2.1.5 from Section 1 in Chapter 2, "The Pigeonhole Principle".

(b) *Answer*: the agent cannot be trusted.

Solution. Suppose, on the contrary, that each of the 15 countries participated in the conclusion of three bilateral treaties. The total number of treaties is 15/2, since each treaty involves two countries. But this number is not an integer. Contradiction.

(c) *Answer*: yes.

3.1.4. *Answer*: yes.

Suggestion (Written by M. Skopenkov). Put a chip in an arbitrary triangle. At each step, move it to the neighboring triangle, but not to the one from which it just came. The chip will never be able to return to any triangle for a second time; at each step it intersects a certain diagonal, and the movement continues in another half-plane relative to this diagonal. The triangle from which further movement of the chip is impossible is one of the desired ones. Now we repeat the process, starting from the triangle found.

3.1.5. Construct a graph as follows. The vertices of the graph correspond to members of our company. Connect the vertices by a red edge if corresponding members know each other, and with a blue edge otherwise. Note that any pair of vertices will be connected by an edge. Therefore, from each vertex of this graph exactly five edges come out.

Therefore, for each vertex at least three edges AB, AC, AD have the same color. Let is assume that these three edges are red. (It is clear how to modify the arguments if there are three blue edges.)

If one of the edges BC, CD, and BD, for example BC, is also red, then the triangle ABC is red. If all the edges BC, CD, and BD are blue, then the triangle BCD is blue.

So, in any case there is a triangle with sides of the same color.

3.1.7. *Solution* (Written by A. Kutsenko). (a) We denote the set of deputies by D. Let A be the largest (by size) subset of the set D in which no deputy threatens another.

Assume, contrary to what we need to prove, that $a = |A| < 150$. Note that each deputy from A threaten some deputy from $D - A$. We denote by X the set of deputies from $D - A$ who are not threatened by anyone from A. The set of $D - A$ contains $450 - a$ elements, while no more than a deputies from $D - A$ were threatened by someone from A. Therefore $|X| \geq 450 - 2a > 150$.

Suppose a deputy from the set X does not threaten any deputy from the set A. Then this deputy can be added to the set A and it will become larger (while maintaining the condition that no one threatens anyone in it), which contradicts the maximality of the set A. This means that there is no such deputy, and each deputy from X threatens some deputy from A.

In particular, this means that deputies from X do not threaten each other. But we know that $|X| > 150 > |A|$, and this contradicts the maximality of the set A.

(b) We will argue similarly to part (a). Let D be the set of all gangsters. We will use contradiction. Suppose that any subset of gangsters where no one is hunting for anyone is finite. Now let A be the maximal (by inclusion) subset of gangsters in which no gangster hunts for any other from A. (It's not so easy to justify why such a set A exists, and we won't do it. To do this, we would need the *Axiom of Choice*, one of the most subtle axioms of set theory.)

Denote by X the set of gangsters from $D - A$ such that no one from A hunts any of them. Since D is infinite and A is finite (in particular, gangsters from A hunt for a finite number of gangsters), X is infinite. Suppose there are gangsters in X who hunt for no one from A. But then these gangsters can be added to A (while maintaining the condition that no one is hunting for anyone in it), which contradicts the maximality of A. So every gangster from X is hunting for someone from A. In particular, this means that gangsters from the set X do not hunt for each other. But X is infinite, a contradiction. So, there exists an infinite set of gangsters not hunting each other.

3.1.8. Let us call "even" those streets along which the tourist walked an even number of times on the way to the cafe; the rest we call "odd". On her way to the cafe the tourist left the station one time more than she returned, which means that she took the streets coming out of the station an odd number of times. Therefore, an odd number of odd streets are coming out of the station. Similarly, an odd number of odd streets also are coming out of the cafe, and there is an even number of odd streets from any other intersection.

Let the tourist go from the cafe along an arbitrary odd street passing each street at most once. When she enters an intersection other than the train station, there remains an odd number of unused odd streets coming out of this intersection. Therefore, she can leave this intersection. So, at the moment when the tourist has nowhere to go, she is already at the station.

3.1.9. Consider the person A with the maximum number of acquaintances. Let his friends be $B_1, B_2, \ldots, B_n$. Every two people B_i and B_j have a common friend A, so B_i and B_j do not know each other. B_1 and B_i $(i = 2, 3, \ldots, n)$ have a common friend C_i different from A. If C_i and C_j for some $i \neq j$ are the same person, then C_i and A have common friends B_1, B_i, and B_j, which is impossible by hypothesis. So B_1 has at least n acquaintances: $A, C_2, C_3, \ldots, C_n$. Due to the maximality of n, B_1 has exactly n acquaintances. Similarly, all B_i, $i = 2, 3, \ldots, n$, have exactly n acquaintances, from which, in turn, it follows that all friends of B_i also have n acquaintances. Each person is familiar with either A or one of the B_i's, which means that everyone in the society has exactly n acquaintances.

2. Counting in graphs (2)

3.2.1. A person is called noncommunicative if she has less than 10 friends. A person is called eccentric if all her friends are noncommunicative. Prove that the number of eccentric people does not exceed the number of noncommunicative people.

3.2.2. Ten people came to a party. Those who do not know any of the other people leave. Then those who know exactly one person among those remaining also leave. Then, similarly, those who know exactly $2, 3, 4, \ldots, 9$ among those remaining at the time leave. What is the maximal number of people that can remain at the end?

3.2.3. There are 2007 teams that participated in a tournament where every team plays every other and there are no draws. Find the maximal possible number of triples of teams such that the first team defeated the second, the second defeated the third, and the third defeated the first.

3.2.4. Some pairs of cities are connected by (nonstop) flights. Cities connected by a flight are called *adjacent*. Could there be exactly 100 flights in this country if for any two adjacent cities there is exactly
  (a) one city adjacent to both of them;
  (b) two cities adjacent to both of them?

3.2.5. A company employs 50 000 people. For each of them, the sum of the number of her immediate supervisors and her immediate subordinates is 7. On Monday, each employee issues an order and gives a copy of this order to each of her immediate subordinates (if any). Further, every day the employee takes all the orders she received in the previous days and either distributes copies of them to all her immediate subordinates or, if she doesn't have any, carries out the orders herself. It turned out that on Friday no papers in the company have been distributed. Prove that the company has at least 97 employees who have no supervisors.

3.2.6.* In each of the three schools, there are n students. Any student has exactly $n + 1$ friends among students from two other schools. Prove that you can choose one student from each school so that each of the three selected students is a friend of the other two.

Suggestion. To solve the problems of this section, you need not study the structure of the company, the flight network, or the tournament, but just count the *number* of some objects: couples (manager, subordinate), couples of adjacent cities, couples of acquaintances, triples of acquaintances, etc.

Suggestions, solutions, and answers

3.2.1. We will call noncommunicative people who are not eccentrics "just noncommunicative", and eccentrics who are not noncommunicative, we will call "just eccentrics". Noncommunicative eccentrics cannot be familiar with just eccentrics, which means that just eccentrics are familiar only with just noncommunicative people. Everyone who is just eccentric is familiar with at least 10 just noncommunicative people, and every noncommunicative person is familiar with at most 9 just eccentrics.

Let there be n acquaintances between just eccentrics and just noncommunicative people. The number of just eccentrics is at most $n/10$, since the number of just noncommunicative people is at least $n/9$. We see that there are at most as many just eccentrics as just noncommunicative people, which means that the total number of eccentrics cannot exceed the total number of noncommunicative people.

3.2.2. *Answer*: 8.

Suggestion. It is easy to verify that if all who came, except for two people A and B, are familiar with each other, then at the end all except A and B stay; i.e., 8 people stay. We prove that 9 people could not stay. It is clear that the person A, who initially had the smallest number of friends (k), will leave at some point. If no one else leaves, then everyone else (except A) had more than k acquaintances before A left and less than $k + 1$ after he left. But then A should be familiar with everyone else, i.e., $k = 9$, which contradicts the strict minimality of k.

3.2.4. (a) *Answer*: no.

Hint. In such a graph, the number of edges E is divisible by 3.

(b) *Answer*: no.

Hint. In such a graph, the number of edges E is divisible by 3. Indeed, we calculate in two ways the total number of pairs (the edge and the triangle containing it). Since there are two triangles adjacent to each edge, then the number of such pairs is $2E$. Since each triangle has three edges, the amount of $2E$ is equal to three times the number of triangles. So E is divisible by 3.

3.2.5. *Solution* (from the site http://problems.ru). If the company has k supreme bosses (i.e., employees with no supervisors), then every employee should finally see at least one of the k orders issued by these bosses on Monday. On Monday no more than $7k$ workers saw them, on Tuesday there are no more than $7k \cdot 6$, on Wednesday there are no more than $7k \cdot 36$ workers. Everyone who saw these orders on Thursday has no subordinates; that means they all have 7 supervisors each. These supevisors could see one of k orders only on Wednesday. Therefore, the number of these supevisors does not exceed $7k \cdot 36$, and each of these supervisors has at most 6 subordinates. Thus, on Thursday, orders have been seen by at most $(7k \cdot 36) \cdot 6/7 = 6k \cdot 36$ employees. Therefore $50\,000 \leq k + 7k + 42k + 252k + 216k = 518k$, which means that $k \geq 97$.

3. Paths in graphs (2)

A *graph* can be thought of as a set of points (for example, on a plane), some pairs of which are connected by broken lines.[1] The points are called *vertices* of the graph, and the broken lines are called *edges*. Only the endpoints of each broken line are the vertices of the graph; broken lines can intersect at nonendpoints, but such intersection points are not considered as vertices.

We assume that each edge connects different vertices and each pair of vertices is not connected by more than one edge. The common name for such graphs is *graphs without loops and multiple edges* or *simple graphs*.

A common example of a graph is a dating graph. The vertices of this graph correspond to people; two vertices are connected by an edge if the corresponding two people are dating.

The number of vertices and the number of edges of the graph in question are denoted by V and E, respectively.

The *degree* of a vertex of a graph is the number of edges coming out of it.

3.3.1. *Degrees of vertices.* (a) In any graph there are always two vertices of the same degree.

(b) The sum of the degrees of the vertices in any graph is even and is equal to $2E$.

A *path* in a graph is a sequence of vertices such that any two consecutive vertices are connected by an edge. A *cycle* is a path in which the first and last vertices coincide. A path (cycle) is called *nonself-intersecting* (or *simple*) if it passes through each vertex at most once and does not pass any edge twice in a row.

A graph is called *connected* if any two of its vertices can be connected by a path.

3.3.2. Suppose we are given a graph with degree of any vertex at least k, where $k \geq 2$. Prove that in this graph there is a simple cycle of length not less than $k + 1$.

3.3.3. Let a 2-connected graph, i.e., a connected graph which remains connected if any of its edges is removed, be given. Two players take turns

---

[1] A formal definition of a graph, path, and cycle is given, for example, in [**GDI2**]. The presented problems can be solved using the informal definitions given here. In this section, only *undirected graphs* are considered.

putting arrows on the edges. The player loses if after her move it becomes impossible to go from some vertex to some other vertex, moving only along arrows in the direction they point at and along edges without arrows. Prove that if both players choose optimal moves the game will end in a draw.

3.3.4. (a) In the city there are no bridges, no tunnels, no dead ends. All intersections are formed by the intersection of exactly two streets (streets are not necessarily straight). When making an inspection tour of the city, the governor, at each intersection, instructs the driver to turn either right or left. After some time the governor's driver noticed that they were driving along the road along which they had already driven. Prove that they are traveling along this road in the same direction as they did the first time.

(b)* Two climbers stand at sea level on opposite sides of a ridge (broken line with a finite number of sides) entirely above sea level. Prove that they can meet staying all the time at the same height above sea level.

The distance between two vertices of a connected graph is the minimal number of edges in the path connecting them (the minimum is taken over all paths connecting these vertices). The distance in the graph satisfies the triangle inequality. *The diameter* of a connected graph is the largest distance between vertices of the graph.

3.3.5.* Suppose that in a connected graph of diameter d the minimal cycle length is $2d + 1$. Prove that the degrees of all vertices are equal.

Suggestions, solutions, and answers

3.3.1. (a), (b) These problems are reformulations of the problems 3.1.3(a), (b).

3.3.2 (Written by M. Skopenkov). Put the chip on an arbitrary vertex. At each step, we will move it to the neighboring vertex, which it has not visited yet. Consider a position from which further movement of the chip is impossible.

3.3.4 (Written by M. Skopenkov). (a) Color the areas in a checkerboard pattern. Then the governor goes around the black areas clockwise, and around the white ones counterclockwise.

(b) Let the ridge be the graph of the function $f(x)$. Consider the graph $\{(x, y) : f(x) = f(y)\}$. See also §6 in Chapter 7, "Phase Spaces".

3.3.5. See solution in the article [**Kru**].

Chapter 4

Constructions and invariants

This topic is also appropriate for students of grades 6 and 7, but they need to use not this chapter, but the article [**LT**] and the corresponding chapter of the book [**FGI**].

1. Constructions[1] (1) *By A. V. Shapovalov*

If, when asked the question "Can it happen?", you suspect that the answer is "It can", then it is worth asking yourself: "How can this happen?" Clarify the question: "What properties should this construction have?" Additional knowledge will greatly narrow your search. Ask questions throughout the construction. You will be surprised to see how many constructions turn out to be logical and the only ones possible.

Often there are many examples, but only one is needed. An excess of freedom can be confusing: it is not clear where to start. Apply *common sense* and *natural considerations*. They limit the field to search for an example, but then the search speeds up and is easier. Generally your experience is much bigger than you think. The answer may be *a well-known object*; you just need to look at it at the right angle.

4.1.1. Two triangles have two sides equal, and also the altitudes drawn to their third sides are equal. Are these triangles necessarily congruent?

4.1.2. Is it true that positive numbers can be put at the vertices of an arbitrary triangle so that the length of each side is equal to the sum of the numbers at its endpoints?

4.1.3. In a math circle, each participant has exactly 6 friends. Can each pair of participants have exactly two mutual friends?

A construction with a large number of parts is easier to build from identical "bricks". Even if they all cannot be the same, try to take more

---

[1]This collection of problems is based on the books [**Sha14**] and [**Sha15**].

similar ones. You can still choose two types of parts and calculate how many parts of each types are needed.

If the parts "for assembly" are given and they are distinct, then it is worth trying to combine these parts into *identical blocks* and to build from these blocks.

4.1.4. A nonnegative integer is called a *zebra* if its digits alternate in parity (even and odd) and there are at least three different digits. Can the difference of two 100-digit zebras be a 100-digit zebra?

4.1.5. The faces of the parallelepiped with edges 3, 4, and 5 are divided into unit squares. In each square was entered a natural number. Consider all possible rings one square wide, parallel to some face. Can the sums of the numbers in the squares in each such ring be the same?

In problems where equal parts are required, you have to choose a form for the parts. This might help: the parts are obviously equal if they are obtained from each other by symmetry, translation, or rotation For example, in the case of the square, cuts preserved by a rotation through 90° are popular, and in the case of a regular triangle, the cuts preserved by a rotation through 120° are popular. For symmetric objects the search of an example starts with symmetrical or "almost symmetrical" constructions. Symmetry and the idea of "arranging objects in a circle" are also applicable in nongeometrical problems.

4.1.6. Is it possible to number the edges of a cube with numbers -6, -5, -4, -3, -2, -1, 1, 2, 3, 4, 5, 6 so that for each triple of edges coming from a vertex, the sum would be the same?

4.1.7. The circle was cut into several congruent parts. Does the boundary of each part necessarily pass through the center of the circle?

If conflicting requirements are imposed on the design, take a closer look. Often these contradictions are imaginary. E.g., *large* perimeter does not contradict *small* area. In general, for the words like "a lot" and "a little", you need to be able to give exact mathematical meaning in the solution using equalities and inequalities.

4.1.8.* An iceberg having the form of a convex polytope floats in the sea. Can it happens that 90% of its volume is below the water level and at the same time more than half of its surface is above the water level?

4.1.9. Let A and B be two cubic dice with nonstandard sets of numbers on the faces. Let $t(A)$ be the result of throwing A, and let $t(B)$ be the result of throwing B. We throw dice independently of one another and say that the die A *wins* over the die B if the probability that $t(A) > t(B)$ is greater than the probability that $t(B) > t(A)$. Can there be three dice A, B, and C such that A wins over B, B wins over C, and C wins over A?

Here is an equivalent definition of when A wins over B that does not use probability. For a pair of dice A and B we have 36 ordered pairs (number on a face of A, number on a face of B). Die A *wins* against die B if the number of pairs where the first number is greater than the second exceeds the number of pairs where the second number is greater than the first.

Invisible barriers in the solver's head can interfere with solving the problem. If the obvious solution is not visible, you need to expand the list of options until it is complete (if possible). *Inertia of thinking* manifests itself in the fact that one might miss the key option or might not suspect that there is more than one option. Use the "Sherlock Holmes Method": "When you have eliminated all that is impossible, whatever remains, however improbable, must be the truth."

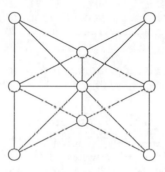

FIGURE 1

4.1.10. There are 9 apples on the table, forming 10 rows of 3 apples in each (see Fig. 1). Nine rows are known to have the same weight, but the weight of the tenth row is different. There are electronic scales on which for 1 dollar you can find out the weight of any group of apples. What is the least amount of money that must be paid to find out the weight of the row that is different?

4.1.11. Can a line break a hexagon into four congruent triangles?

Reduction is the procedure that reduces a complicated problem to a simpler one. So, if a complex structure cannot be built immediately, construct its *necessary part*. Even if this part cannot be completed to the whole, solving a simplified problem can serve as a warm-up, after which you will return to the difficult problem already with accumulated experience.

4.1.12. Baron Munchausen says he has a multidigit *palindrome* (that is, a number that reads the same from left to right and from right to left). Having written it down on a paper tape, the baron made several cuts between digits. The tape broke into N pieces. By arranging pieces in a different order, the baron saw that the numbers 1, 2, ..., N were recorded on the pieces (each number exactly once). Can the baron's claim be true?

When creating a construction, ambiguity of choice may interfere. In a *bottleneck*, everything is clear or uncertainty is minimal, which reduces the brute-force search. Starting from a bottleneck, we will either quickly arrive at a contradiction or we will build a large piece of construction. How do we look for bottlenecks? Take a closer look: they serve as obstacles to the construction of the structure or they seem to be such.

4.1.13. Having written down the numbers 1, $\frac{1}{2}$, $\frac{1}{3}$, ..., $\frac{1}{10}$ in some order, connect them with the four arithmetic signs so that the resulting expression is equal to 0. (Parentheses cannot be used.)

4.1.14. Do there exist three congruent heptagons, all of whose vertices match but none of whose sides match?

4.1.15. Is it possible to cut a triangle into four convex figures: triangle, quadrilateral, pentagon, and hexagon?

Creating a *gradual construction* one goes step by step through auxiliary constructions — *blanks*. On each step, the construction *is improved* to the next. In the workpiece, the requirements for the final design are fulfilled only partially. We keep *principle* conditions but temporarily forget or weaken *technical* ones.

4.1.16. Can all sides and all altitudes of an acute triangle have integer length?

4.1.17. Prove that there exists a palindrome divisible by 6^{100}. (Recall that a *palindrome* is a number that does not change when writing its digits in reverse order.)

Finally, with *construction by induction*, the result is obtained gradually, but in an infinite number of steps. Section 5 of Chapter 2, titled "Finite and countable sets", is devoted to such constructions.

Continue your acquaintance with constructions using article [**GK**] and books [**Sha14, Sha15, Sha08**].

Suggestions, solutions, and answers

Below, problem solutions and *paths to solution* are carefully separated. A solution is what the problem solver ideally should write. A path to the solution must remain in the heading; here it explains how a solution could be devised. In construction problems, a solution and a path to a solution usually have little in common.

The solution of a construction problem consists of two parts: *an example*, that is, a description of the construction, and *proof* that the construction satisfies the condition of the problem. For our problems, the second part is not difficult and is usually omitted. But sometimes from many possible examples one still needs to choose one for which the proof is simpler.

4.1.1. *Answer*: not necessarily.

Solution. Consider the isosceles triangle ACD and the point B on the extension of the base DC. In triangles ABC and ABD, the sides AB and altitude AH are common, and the sides AC and AD are equal. However, these triangles are not equal: one is a part of the other.

Path to solution. Let's try to *construct* a triangle with two sides b, c and the altitude h drawn to the third side. To do this, draw a line l (a third side will lie on it) and build the vertex A at a distance of h from l. Two other vertices of the triangle must lie on this line at distances b and c from point A. Drawing circles of the indicated radii with center at point A, we obtain (for $b > h$ and $c > h$) two points of intersection of each of the circles with l. We see that, up to symmetry, there are two fundamentally different triangles: when the vertices are selected on one side of the point of the line nearest to A and on different sides of it.

4.1.2. *Answer*: it is true.

Solution. We inscribe a circle in the triangle. Segments of tangents from each vertex to the circle are equal. We write down the length of such a segment at the vertex. Since each side consists of two such segments, the sums are at its endpoints and will be equal to its length.

Path to solution. It is natural for a geometer to look not just at numbers, but at the lengths of the segments. This *reduces* the problem to the following: break up each side into two smaller segments so that for each vertex the lengths of two adjacent segments are equal. Pairs of equal segments in geometry are not rare, but the solution for three pairs of equal segments suggests itself: these, of course, are the segments of the tangents to the inscribed circle.

But you can also solve this problem algebraically: compose a system of linear equations for the desired numbers, solve it, and check positivity of solutions with the help of the triangle inequality.

4.1.3. *Answer*: they can.

Solution. Let 16 people stand in the cells of a 4×4 square, and assume that everyone is a friend only with people in his column and his row.

4.1.4. *Answer*: it can.

Solution. For example, $50945094\ldots5094-25472547\ldots2547 = 25472547$ $\ldots2547$.

Path to solution. Let's look for a zebra equal to the sum of two zebras. It will be simpler to take equal terms. For two-digit numbers, there are such examples: $25 + 25 = 50$, $47 + 47 = 94$. (Of course, for two-digit numbers we do not require three different digits.) From these blocks we can assemble a 100-digit example.

4.1.5. *Answer*: they can.

Solution. In all squares of a 3×4 face we write the number 5, in squares of a 3×5 face we write 8, and in all squares of a 4×5 face we write 9.

Path to solution. Let us look for a solution where in all squares of a 3×5 face the same number x is written, in squares of a 4×5 face y is written, and in squares of a 3×4 face z is written. Comparing the sums in rings parallel to different faces, we obtain the system of equations $5x+4z = 3z+5y = 4y+3x$. One of the solutions of this system is $(8, 9, 5)$.

4.1.6. *Answer*: it is possible.

Solution. See Fig. 2.

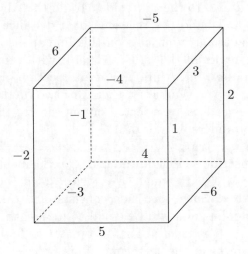

FIGURE 2

Path to solution. Adding sums at all vertices, we get twice the sum of numbers at all edges, i.e., 0. Hence, the sum at the vertices is zero. The endpoints of edges with opposite numbers cannot coincide; otherwise the sum at such an endpoint is not 0. Also endpoints of edges with numbers 6 and -6 cannot be connected by an edge because otherwise in one of its endpoints, the sum is not 0. This leads to the following idea: one should search for a *symmetric* solution by placing opposite numbers at opposite edges.

4.1.7. *Answer*: not necessary.

Solution. See Fig. 3. First we divide the circle into six equal arcs; then using the same arcs we connect the division points with the center of the

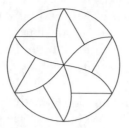

FIGURE 3

circle. We get six equal symmetrical parts. Divide each part into two equal smaller parts with a line segment.

Path to solution. A *bottleneck* is a piece of the boundary of a part that is congruent to an arc of the original circle. It is clear that each part has such pieces. If all of them are parts of the original circles, then most likely all parts are obtained from each other by turning around the center. At least one of the parts contains the center, so each of them will contain the center. So there should be arcs on the boundary of each piece that do not lie on the original circle. Connecting equidistant points with the center of a circle by such arcs we split a circle into n equal parts for any n. Every part is bounded, among other things, by two convex arcs; the one going to the center is $1/6$ of the circle, and the other one is $1/n$. If we want to cut further, it would be worth making these arcs equal. And indeed, when divided into 6 parts, the figure turns out to be symmetrical, and it can be divided into two equal parts by a segment.

4.1.8. *Answer*: it can.

Example. A regular pyramid floating with its apex down. The base of the pyramid is parallel to the surface of water and is located so that 90% of the volume is in the water. The angle of inclination of the lateral faces to the base is taken to be less than $\arccos(2\sqrt[3]{0.9^2} - 1)$.

Proof that the given example satisfies the condition. Let the area of the side surface of the pyramid be equal to S. Then the area of the base is $S \cdot \cos\alpha$, where α is the angle of inclination of the lateral faces to the base. By construction, part of the pyramid in the water is the pyramid similar to the original one with the similarity coefficient $k = \sqrt[3]{0.9}$. The area of the side surface of this part is $k^2 S$. By construction, $\cos\alpha > 2k^2 - 1$ (such angle α exists because $2k^2 - 1 < 1$). But then $2k^2 S < S + S \cdot \cos\alpha$; that is, less than half the surface of the pyramid is in the water.

Path to solution. At first sight, this is impossible; after all, only a small hat of a spherical iceberg is above the water. However, the cube will have a larger area since it does not get fatter in the middle like a ball. Now, why should an iceberg *expand* when going down, or stay the same? Why can it not get *narrower going down*?! Take the inverted pyramid. Only a flat piece near the base will stick out above the water level, and most of the side faces will be under water. Why most? Well, because the lateral surface is larger than the base area, and much larger, and 90% (or so) of it is in the water. And why, in fact, most? Indeed, if you flatten the pyramid, then the side surface will be close to the area of the base, but part of it is still above the water! And well, let's calculate this carefully....

4.1.9. *Answer*: it can.

Solution. Suppose, for example, that die A has the number 3 on all faces, the die B has 2, 2, 2, 2, 5, 5, and the die C has 1, 1, 4, 4, 4, 4. When comparing the dice A and B, the pair $(3, 2)$ will appear 24 times and the pair $(3, 5)$ will appear 12 times. When comparing the dice B and C, the winning pairs for B are the pair $(2, 1)$ (appears 8 times), the pair $(5, 1)$ (appears 4 times), and the pair $(5, 4)$ (appears 8 times), for a total of 20 wins, which is more than half. Finally, when comparing dice C and A, the pair $(4, 3)$ will appear 24 times, so C wins over A.

Path to solution. It is clear that a die with a big sum of points on all faces has an advantage; therefore it is *more convenient* to look for an example where the sums on all dice are the same. A brute-force search of options is shorter if among the numbers on faces many are equal. Finally, the following idea helps to construct an example: "Since the number of wins is more important than the point difference of each particular win, try to win a little and to lose a lot."

4.1.10. *Answer*: 0 dollars.

Solution. The sum V of weights of the three vertical rows is equal to the sum of weights of all apples. The sum D of the weights of the six diagonal rows is equal to twice the sum of the weights of all apples. So $2V = D$. If in this equality 8 of the nine terms are equal, then the ninth is equal to each of them. So, the horizontal row is different.

Comment. Deviation of one row is possible; for example, see Fig. 4.

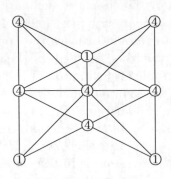

FIGURE 4

4.1.11. *Answer*: it can.

Solution. See Fig. 5. The line divides the hexagon into 4 right triangles with legs m and n.

Path to solution. It seems unbelievable even that *one* line can split a hexagon into *four* triangles. For starters, it is good to understand *how* in general *this can happen*! But if it can, then this is an explicit *bottleneck*. In triangles, one side belongs to the section, but the rest should be sides of the hexagon. There are already 8 sides! However, the sides of the triangles can also be *parts* of the sides of the hexagon. So, the cut intersects at least two sides of the hexagon.

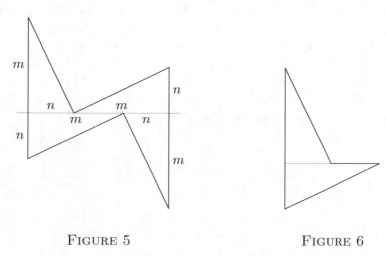

FIGURE 5   FIGURE 6

In this case, pairs of angles with the sum of 180° are formed. Now we use that the triangles are congruent. If the angles in the pair are not equal, then in one of the congruent triangles there is a pair of angles with the sum of 180°, which is impossible. This means that the angles are equal; that is, the cut intersects the sides at right angles. So, four right-angled triangles adjoin legs to the section in pairs; the vertices with right angles are pairwise equal. It remains to see that the adjacent legs of the triangles are not equal; otherwise the pair would form an isosceles triangle, which would be connected to the rest of the hexagon by one vertex. This means that each pair of triangles forms a quadrilateral, as in Fig. 6, and by gluing two such quadrilaterals along the side that extends the cut, we obtain the desired hexagon.

4.1.12. *Answer*: they can.

Solution. Here is an example for $N = 19$ (points mark the location of the cut):

$$9.18.7.16.5.14.3.12.1.10.11.2.13.4.15.6.17.8.19.$$

Path to solution. In a palindrome, all digits (except, perhaps, the middle one) can be divided into pairs of equal ones. Therefore, at most one digit appears an odd number of times, and all other digits must appear an even number of times. We will look for the minimum N such that for a set of digits in writing the numbers $1, 2, \ldots, N$ this *parity condition* is satisfied. However, if $N = 2, \ldots, 9$, then the numbers $1, 2, \ldots$ occur once in a set, and if $N = 10, \ldots, 18$, then the numbers 0 and 9 occur once. So, the smallest N

for which the parity condition is satisfied is $N = 19$; in the record $1, 2, \ldots, 19$ the digit 0 occurs 1 time, the digit 1 occurs 12 times, the rest of the digits occur 2 times each. Now, starting from the middle, you can construct a palindrome.

4.1.13. *Answer*: for example, $\frac{1}{5} \div \frac{1}{10} \div \frac{1}{7} - \frac{1}{3} \div \frac{1}{9} \div \frac{1}{2} - 1 \div \frac{1}{6} - \frac{1}{4} \div \frac{1}{8}$.

Path to solution. The bottleneck is the fraction $\frac{1}{7}$. We must divide by it; otherwise it, and possibly its multiple, forms a single fraction with a denominator that is a multiple of 7. Then, when reduced to a common denominator, the denominator must be a multiple of 7, and all but one of the numerators is a multiple of 7. As a result, the algebraic sum will be a fraction which is irreducible by 7 and, therefore, will not be an integer.

4.1.14. *Answer*: they do.

Solution. In Fig. 7 the solid, dashed, and dot heptagons are obtained from each other by rotations around a central point.

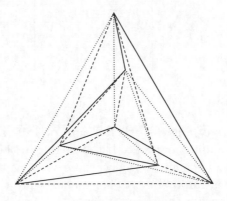

FIGURE 7

Path to solution. There are many bottlenecks in this design. In three heptagons there are 21 sides. Between seven points there are $\binom{7}{2} = 21$ segments. So, each segment should become the side of a heptagon.

Consider the convex hull of 7 vertices. Suppose that at least 4 of them are on the boundary. Then some diagonal AB of the convex hull separates some two vertices C and D. As shown above, the segment AB will be a side of one of the heptagons, and C and D will be vertices of this heptagon. Since C and D lie on opposite sides of AB and the heptagon lies entirely inside the hull, it intersects itself. We get a contradiction. Therefore, the convex hull is a triangle.

The repetition of the triple (three vertices, three heptagons) suggests looking for an example where heptagons transform into each other with a rotation by a third of a full turn. Place three vertices at the vertices of a regular triangle, one in its center, and three more at the vertices of a smaller regular triangle with the same center. Further it remains to color the edges in three colors so that edges of each color make a heptagon. It's more convenient to preset how colors turn into each other when turned clockwise and to paint edges turning into each other by triples.

4.1.15. *Answer*: yes.

Solution. For example, see Fig. 8.

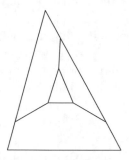

FIGURE 8

Path to solution. The first bottleneck will be the contact of parts between themselves and with the sides of the original triangle. Indeed, due to the convexity, each of the parts is in contact with no more than three other parts. So the remaining sides of a part must lie on the sides of the original triangle. There are three such sides for the hexagon, at least two for the pentagon, and at least one for the quadrilateral.

Second bottleneck: this is the connection of parts to the angles. It is important to understand that removal of any part should not lead to a breakdown of the original triangle into disconnected parts. This is obvious for nonhexagonal parts; all other parts are connected through the hexagon. Well, if the original triangle falls apart after removing the hexagon, then the pentagon touches no more than two sides and not more than two other parts, which is a contradiction. From this property it follows that the adjacency of the parts to the original triangle can be only like this: two vertices of the original triangle belong to the hexagon, and the third belongs to the pentagon. All further adjacencies and the main picture are determined unambiguously.

From the same considerations, we can prove that a triangle cannot be cut into more than 4 convex polygons with different numbers of sides (see [**Z**]).

4.1.16. *Answer*: they can.

Solution. For example, in an isosceles triangle with base 30 cm and sides 25. The altitude dropped to the base divides it into two right triangles, and by the Pythagorean theorem it is 20. In these triangles, the angle against the leg of length 15 is less than $45°$; therefore the angle at the apex of the isosceles triangle is acute. Its area is $S = 300$. So, the altitude dropped to the side is $2 \cdot 300/25 = 24$.

Path to solution. Weaken the condition: let the triangle be acute or right. Among the right triangles with integer sides, *Egyptian* is the most well known with legs 3 and 4 and hypotenuse 5. There are already two heights that are integers, but the third height is not an integer; it is $3 \cdot 4/5 = 2.4$. We use this example as a workpiece. If all sides are increased proportionally by 10 times, then not only will the sides remain integers, but also the height will become integer. The essential thing in the workpiece was that both sides and the area were integers. But then it is easy to make an acute triangle out of two right ones.

4.1.17. *Solution.* We show first that there is a palindrome Π, divisible by 2^{100}. Let A be the number obtained by writing the digits of 2^{100} in reverse order. We write down the digits of the number A, then 100 zeros, and, finally, we write 2^{100} on the right. This gives us a palindrome. It is equal to the sum of a number that ends in more than 100 zeros and the number 2^{100}. Both terms are divisible by 2^{100}; therefore their sum is also divisible by 2^{100}.

Write the palindrome Π three times in a row. Get a palindrome Π_1, divisible by 3. Then we write the palindrome Π_1 three times in a row. We get that the palindrome Π_2, divisible by Π_1, with the quotient written using three units and several zeros, is divisible by 3, so Π_2 is divisible by 9. Continuing similarly, after 100 steps we get the palindrome Π_{100}, divisible by 3^{100}. It remains to notice that all palindromes built along the way are also divisible by Π, which means that they are also divisible by 2^{100}.

2. Invariants I (1)
By A. Ya. Kanel-Belov

When solving various classes of problems, it often helps to apply a very useful trick, based on the observation that during some transformations a certain quantity does not change, i.e., *is invariant.*

Here is a simple illustration.

4.2.1. The cells of the square $n \times n$ table have plus signs and one minus sign, as shown in Fig. 9. One can simultaneously change the sign in all cells located in one row or in one column. Prove that no matter how much we carry out such sign changes, we will not be able to get a table consisting of just plus signs (a) for $n = 4$ and (b) for $n = 5$.

+	−	+	+
+	+	+	+
+	+	+	+
+	+	+	+

FIGURE 9

4.2.2. The circle is divided into 6 sectors, in which the numbers $0, 0, 1, 0, 1, 0$ are written. One can simultaneously add ones to the numbers standing in two neighboring sectors. Is it possible by several such steps to make the numbers in all sectors equal?

4.2.3. Each vertex of the cube contains a number. In one step increase both numbers placed at (any) one edge by one. Is it possible in several such steps to make all eight numbers equal, if the numbers were originally placed, as shown (a) in Fig. 10 and (b) in Fig. 11?

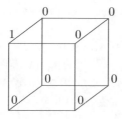

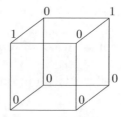

FIGURE 10   FIGURE 11

In the previous problems, the invariant was used to prove the impossibility of moving from one position to another. Sometimes an invariant is used differently, as a means of avoiding a brute-force search.

4.2.4. Two people play on a 1×30 strip. At the beginning of the game, at one edge of the strip stands a white checker, at the other stands the black one. They go in turn, each with the checker of their own color. In one move, the checker moves by one square forward or backward if this square is not occupied by the other checker. The player that cannot make a move loses. Prove that the first player cannot win in this game regardless of the moves she makes.

4.2.5. On the blackboard 10 pluses and 15 minuses are written. One can erase any two signs and write instead of them a plus sign if the same signs are erased, and a minus sign if different signs are erased. What sign will remain last on the board and why does it not depend on the order in which the signs are erased?

4.2.6. Several zeros, ones, and twos are written on the blackboard. One can delete two different numbers and write the third number instead (instead of 0 and 1 write 2 and so on). Prove that if in the result of several erasures only one number is left, then it does not depend on the order in which the erasures were made.

In the previous problems, we made sure that for the proof of the inability to move from one position to another it suffices to present an invariant taking different values in these positions. However, it is clear that if the invariant you constructed gets the same value in two positions, this does not prove the possibility of transition from one position to another, as some obstacles may occur.

4.2.7. Computer BK-0000 has only 2 memory cells, each of which may store a single integer. The computer language includes 3 operations: operation $P: (a, b) \rightarrow (a+b, b)$; operation $M: (a, b) \rightarrow (a-b, b)$; operation $S: (a, b) \rightarrow (b, a)$. Initially, the cells contained numbers 93 and 81.

(a) Is it possible after a certain sequence of operations to get the numbers 1 and 0 in cells?

(b) Under what condition can 1 and 0 be obtained from the numbers a and b?

(c) Under what condition can c and d be obtained from the numbers a and b?

Suggestions, solutions, and answers

4.2.1. (a) Replace the character "+" with the number 1 and the character "−" with the number −1. Note that the product of all the numbers in the table does not change after changing the sign of all the numbers in a column or in a row, since they simultaneously change the sign of 4 numbers. In the initial position, this product is equal to −1, and in the table consisting only of pluses the product is +1, which proves the impossibility of transition.

(b) The method from part (a) now fails, since the evenness of the number of minuses is changing. Therefore, we apply the reduction; notice that it is impossible to change exactly one character in the 2×2 "corner" table. It follows that in the larger table only one sign cannot be changed.

4.2.2. *Answer*: not possible.

Suggestion. We enumerate the sectors counterclockwise starting with any of them. Add up the numbers in sectors $1, 3, 5$, and subtract from this sum the numbers in sectors $2, 4, 6$. The obtained value does not change under valid operations (in other words, it is an *invariant*). At the initial moment, it is equal to 2. If the numbers in all sectors became the same, then our value would become 0. So, the numbers cannot become equal.

4.2.3. (a)*Answer*: not possible.

Suggestion. An invariant is the parity of the sum of the numbers in all vertices.

(b)*Answer*: not possible.

Suggestion. Color the vertices of the cube in black and white so that the endpoints of each edge are colored distinctly. An invariant is the sum of the numbers in white vertices minus the sum of the numbers in black vertices.

4.2.4. The distance between the pieces will be measured by the number of free cells between them. Then, after White's move, the distance between the pieces is always odd. If some player does not have a move, then the distance between the checkers is certainly equal to 0. This means that such a situation can arise only after Black moves.

4.2.5. *Answer*: "−".

Suggestion. Replace the character "+" with the number 1 and the character "−" with the number −1. An invariant is the product of all numbers on the board.

4.2.6. *Suggestion* (N. Kodaneva). Let a zeroes, b ones, and c twos be written on the blackboard. The invariants are the parities of $a+b, b+c, c+a$.

4.2.7. (a) *Answer*: no.

Suggestion. The invariant is the divisibility of both numbers by 3.

(b) *Answer*: if and only if a and b are relatively prime.

Suggestion. Use the Euclidean algorithm.

(c) *Answer*: if and only if either $\mathrm{GCD}(a,b) = \mathrm{GCD}(c,d)$ or $(a,b) = (c,d) = (0,0)$.

Suggestion. The invariant is $\mathrm{GCD}(a,b)$.

3. Invariants II (1)[2] *By A. V. Shapovalov*

The word "invariant" means "invariable". It happens that when we change an object, some property of the objects is preserved. Most often, a certain number does not change. For example, when cutting shapes into parts and assembling new shapes, the total area of the parts does not change. This means, for example, that it is impossible to cut a square with side 1 into parts and to assemble from them the equilateral triangle with side 1.5. Indeed, the total area of the parts is 1, and the area of the triangle is greater.

A little more formal: suppose we can change an object using *allowed operations* and suppose that we were able to associate with each object a *quantity* that is preserved under any allowed transformation. Then this quantity is called an *invariant*[3]. However, below we will also see nonnumeric invariants.

In typical problems "on an invariant" you are asked to prove the impossibility of constructing objects. To do this, find an invariant of all objects that *can* be constructed, and verify that for the objects the value of the invariant is different.

But how do we find the invariant? Start with checking some natural ones: sums, products, areas, perimeters, and their combinations. If the object depends on integers, then the remainder of division by some number, most often it is checking for *parity*, can be an invariant.

---

[2]The ideological predecessors of the problems in "Invariants" and "Semi-invariants" collections were, among others, the corresponding sections in [**K-BK08**].

[3]This sentence is not a formal definition of an invariant. But to solve problems, a formal definition of this concept is not necessary.

4.3.1. Prove that the checkered 99×99 square cannot be cut along the borders of cells into rectangles (possibly not congruent), with the perimeter of each rectangle not divisible by 4.

4.3.2. There is a rigid wire border of a square of area $1\,\text{dm}^2$. It was cut into three parts and soldered again. The result is the border of a planar polygon. What is the largest possible area of this polygon?

An important special case: problems where it is necessary to prove that in a given *process* some result is unattainable. Usually the process consists of a step-by-step transition from one position to another. Here it is sufficient to find an invariant associated with the position; then positions with values of the invariant that are different from the value in the initial position will be unattainable. Invariance needs to be checked at each individual step.

4.3.3. One can replace a triple of numbers (a, b, c) with the triple $(a + b - c, a+c-b, b+c-a)$. Is it possible, starting with the triple $(2013, 2014, 2015)$, to get the triple $(2015, 2016, 2017)$, maybe in a different order?

4.3.4. On the table lies a pile of 200 nuts. In one move, one nut is eaten from the pile and the pile is divided into two (not necessarily equal) parts. Then from one pile with more than one nut, one nut is eaten and this pile is divided into two, etc. Is it possible that after several moves all piles will have exactly 5 nuts?

When searching for an invariant, the desired quantity can be calculated not for the entire object, but for a selected subset or several subsets. A practical way to isolate subsets is to use *coloring*; let us say we color the selected subset or paint all the elements in several colors and make computations for each color separately, and then the results of these computations are combined. Section 4 in this chapter, titled "Ways of coloring", is devoted to this method.

4.3.5. Pete writes down the string of numbers $0, 1, 2, \ldots, 10$ in any order. At each move, he increases two adjacent numbers by 1. Can Pete make all numbers equal after several moves?

4.3.6. A central 2×2 square was cut out of a chessboard. Can the remainder be cut into four-cell figures in the shape of the letter "L"?

An invariant might not be a number, but just some kind of nonchanging property. In simple olympiad problems, nonnumeric invariants are most often associated with alternation or inability to destroy an element with some property. This property may be a match, in particular, a *synchronous alternation* of two "beacons".

4.3.7. A chameleon walks on a checkered board like a lame rook; i.e., with each move he moves into one cell adjacent to the side. Once in the next cell, he either paints himself in the color of the cell or paints the cell in his color. A white chameleon is placed on the board consisting of 8×8 black cells. Can he color the board in a checkerboard pattern?

4.3.8. A die is put on a square of a checkerboard, its face coinciding with the square. The die is rolled on the board, turning each time along an edge. At the end, the die returns to the same square standing on the same face. Can the die be turned by $90°$ about the vertical axis?

Is it true that invariants can only be used to prove the impossibility of something? No, not only. To prove some property, for example, we can assume the opposite and then, using invariants, reduce the assumption to a contradiction.

4.3.9. A cardboard triangle is displaced on a plane, each time rolling through its side. After rolling 2015 times, it arrived at the original place. Prove that the triangle is isosceles.

In more complicated mathematical problems, the invariant can be some important property of a mathematical object, such as rationality and irrationality, finiteness, and infinitness. We just used the concept of *orientation* (when we distinguish between the clockwise or counterclockwise orientation). In higher mathematics, polynomial invariants are common. So in the next problem, as an invariant it is more convenient to take not a number, but an *expression*.

4.3.10. Two points are marked on the line: blue on the left and red on the right. At one step, you can add or delete two marked points of the same color if there are no other marked points between them. Is it possible with such operations to ensure that only two marked points are left on the line, but the blue one will be on the right and the red one on the left?

Using an invariant one can often prove that the result of several operations does not depend on their order. To do this, it suffices to introduce a quantity that does not depend on the order, or does not change for intermediate positions, and to show that the final result is uniquely determined by this quantity.

4.3.11. On a $1\,\mathrm{km} \times 1\,\mathrm{km}$ fenced field additional fences are biult dividing the field into rectangular sections $5\,\mathrm{m} \times 20\,\mathrm{m}$ and $6\,\mathrm{m} \times 12\,\mathrm{m}$. What is the total length of fences?

4.3.12. One hundred numbers are written on the board: $1, \frac{1}{2}, \frac{1}{3}, \ldots, \frac{1}{99}$. At every move, Pete erases any two numbers and writes instead of them the ratio of their product to their sum. Prove that the last number on the board is independent of order of erasing, and find this number.

Knowing which positions or constructions are achievable and which ones are not helps to solve the problems of type "estimate + example"; after all, it is easier to prove an estimate for a narrower range of positions, and often an example with given properties is easier to build.

4.3.13. Thirty coins are arranged in a circle, alternating: three in a row lie face up, three lie tails up, three face up, three tails up, etc. If a coin has two distinct neighbors, it can be turned over. What is the largest number of coins you can put face up using such operations?

4.3.14. Behind the Looking Glass, coins of 7, 13, and 25 guineas are used. Alice used several coins for the pie and received for the change two coins more than she paid. What is the lowest possible price of the pie?

Is it possible to use an invariant to prove that some transformation is possible? This is not that simple. Equal values of an invariant do not guarantee such possibility: there can be another reasons for impossibility; for example there can be another invariant taking different values.

4.3.15. Can a paper circle be cut into several parts along straight lines and arcs of circles in such a way that the pieces can be rearranged to make a square of the same area?

In the cases where the coincidence of the values of the invariant guarantees the existence of a transformation, this must be proved separately. By the way, in such cases the invariant is called a *complete invariant*. For example, the invariants indicated in the solutions of problem 4.3.7 and 4.3.13 are complete: any position with a domino or the same number of matches can be obtained. The Bolyai–Gerwien theorem claims that in the problem of *rearranging* polygons (i.e., cutting them into polygonal parts and assembling all the parts), the area of the polygon is the complete invariant.[4]

A *complete system of invariants* is a set of invariants such that the coincidence of the values of *all* invariants of the set guarantees the possibility of construction (transformation). For example, for the reconstruction of polytopes, in addition to the coincidence of volumes, one should require the coincidence of *Dehn's invariant*. Section 9 of Chapter 7, titled "Is it possible to make a cube from a tetrahedron?", is dedicated to this invariant.

---

[4]There is a remarkable story about this theorem in the book [**Bol56a**].

The following problem cannot be solved without building a complete system.

4.3.16. Let F_1, F_2, F_3, ... be a sequence of convex quadrilaterals, where F_{k+1} (for $k = 1, 2, 3, \ldots$) is obtained as follows: F_k is cut along a diagonal, and one of the parts is turned over and glued along the cut to the other part. What is the largest number of different quadrilaterals this sequence may contain? (Polygons are considered different if they cannot be aligned using an isometry.)

A lot has been written about invariants. Some links to problem collections or analyses of individual beautiful problems are given in the list of references.

Suggestions, solutions, and answers

4.3.1. *Solution.* For any cut, the sum of the areas of the parts is 99^2 and is an odd number. Since all areas are integers, there is a part with an odd area. The lengths of the sides of this rectangle are odd, say $2m + 1$ and $2n + 1$, where m and n are integers. Then the perimeter of this part is $4m + 4n + 4$ and is divisible by 4.

4.3.2. *Answer*: $1\,\mathrm{dm}^2$.

Solution. Let the contour be cut at points A, B, C. Then the original square can be divided into 4 parts: the triangle ABC and three polygons cut off by the sides of the triangle from the square (some parts may turn out to be degenerate polygons of area 0; in this case, the arguments will be similar). When drawing up a new contour, the distances between the three solder points D, E, F will still be equal to AB, AC, and BC (because the parts of the contour are rigid). Therefore, the triangles DEF and ABC are congruent. If the cut off polygons do not overlap with $\triangle DEF$, then the area of the new contour is equal to the old, and if they overlap, then it is smaller than the old. Therefore, the maximum area of the new contour is $1\,\mathrm{dm}^2$.

4.3.3. *Answer*: not possible.

Solution. Since $(a + b - c) + (a + c - b) + (b + c - a) = a + b + c$, the sum of the numbers in the set is invariant. But the required set has a larger sum than the original one.

4.3.4. *Answer*: not possible.

Solution. In one move, the number of nuts decreases by 1, and the number of piles increases by 1, so the sum of these numbers is an invariant. Initially, this number is 201. And if we had n piles of 5 nuts each, the sum would be $5n + n = 6n$, but $6n \neq 201$ since 201 is not divisible by 6.

4.3.5. *Answer*: he cannot.

Solution. We color the digits on even places in black, and those on odd places in blue. Subtract the sum of blue numbers from the sum of black numbers. This difference is invariant since both the blue and the black sums increase at each move by the same number. If all numbers became equal, then our difference would become zero. However, initially it is certainly not zero. Indeed, computing the black and the blue sums, we get $0 + 1 + 2 + \cdots + 10 = 55$ is an odd number. But then the difference of the sums is also odd and cannot be equal to 0.

4.3.6. *Answer*: not possible.

Solution. Color the verticals in turn in black and white. Then the L-piece at any position covers an odd number of white cells: 3 or 1. The 60 cells remaning after the central square is cut off must be cut into 15 figures, and these figures include an odd number of white cells. On the other hand, we have that 30, i.e., an even number, white cells are left. Contradiction.

4.3.7. *Answer*: he is not able.

Solution. Note that after each repainting of the cell (except the first one) the color of the chameleon is the same as the color of the cell that he just left. This means that after each cell repainting, the color of the repainted cell coincides with the color of one of the neighboring (by the side) cells. So, in this process, the presence of a one-color domino (a pair of adjacent cells) will be an invariant. But with the checkerboard coloring, there are no such dominoes.

4.3.8. *Answer*: it cannot.

Solution. We color the board in a checkerboard pattern and the vertices of the die in black and white colors so that the endpoints of each edge are of different colors. We can assume that initially the die stands on a white cell and when viewed from above, the upper left vertex is white. Now we note that when we roll the die, the colors of the cell under the die and the upper left vertex change synchronously, so they are always the same. But if the die ends on the same side in the original cell, but with a rotation, these colors will be different, which means that such a position of the die is impossible.

4.3.9. *Solution.* We name the vertices of the triangle ABC clockwise, starting with A. When rolling, the orders ABC and ACB strictly alternate. If the order ABC was in the initial position, then after an odd number of 2015 rolls we will have the ACB order. Since the triangle lies in the same place, its vertices fall into vertices. Suppose a triangle is not isosceles. Then all its angles are different. But since the triangle fell in the same place, each angle covered an angle equal to it; that is, each vertex fell into place. So, the order of the vertices is still the same. Contradiction. Therefore, the triangle must be isosceles.

Comment. In the problem it is claimed that the triangle returned at the same place, so presentating an example of the appropriate isosceles triangle is not required. But here it is for those who are curious: the isosceles triangle with angle $360°/2015$ at the vertex rolls clockwise 2015 times across the side, leaving the vertex in place all the time. Then it will make a full circle.

4.3.10. *Answer*: not possible.

Solution. Moving from left to right, put the letter R under each red dot, and put B under the blue one. Once again moving from left to right, put the signs in front of the letters, starting with plus and strictly alternating pluses and minuses. Considering the record as an expression, we combine the like terms. We note that such an expression will be invariant: after adding or deleting monochromatic points, two opposite terms are added or deleted, and the remaining terms do not change. Since for the initial position our invariant is $B - R$ and for the desired position it should be $B - B$ or $R - B$, a pair with a blue dot on the right is impossible to get.

4.3.11. *Answer*: 248 km.

Solution. The length of the fences is determined by the sum of the perimeters of parcels of lands, and the sum of the perimeters is equal to twice the length of internal fences plus the length of the fence around the field. But is the sum of the perimeters independent of the number of parcels and of their location? The sum of the areas is an invariant, but the perimeter is not the area .... However, let us compare: a $5\,\mathrm{m} \times 20\,\mathrm{m}$ parcel has perimeter $50\,\mathrm{m}$ and area $100\,\mathrm{m}^2$, and a $6\,\mathrm{m} \times 12\,\mathrm{m}$ parcel has perimeter $36\,\mathrm{m}$ and area $72\,\mathrm{m}^2$. Observe that, in both cases the area is equal to the perimeter times $2\,\mathrm{m}$. But then the sum of the perimeters is equal to the sum of the areas divided by $2\,\mathrm{m}$; that is, $1000^2/2 = 500\,000\,\mathrm{m} = 500$ km. Since the length of the fence around the field is 4 km, the total length of the fences is $(500 - 4)/2 = 248$ km.

4.3.12. *Solution.* Rewrite 1 as $\frac{1}{1}$. If Pete erased $\frac{1}{m}$ and $\frac{1}{n}$, then he wrote $\frac{1/m \cdot 1/n}{1/m + 1/n} = \frac{1}{m+n}$. This means that if you write all the numbers as fractions with numerator 1, then the sum of the denominators will be invariant. But then the last remaining number will be equal to $\frac{1}{1+2+\cdots+99} = \frac{1}{4950}$.

4.3.13. *Answer*: 25.

Solution. An *example* is easy to obtain by flipping two extreme coins in each tail's group.

Estimate. We put a match between each two neighboring coins lying differently, 10 matches in total. Note that a coin can be turned only if there is a match at one side of it and at the other there is no match; the match will move to the other side of this coin. Therefore, the number of matches is an invariant. Coins between adjacent matches are all heads or tails; the groups of heads and tails alternate. There are always at least 5 groups of tails; therefore, the number of tails is always at least 5. So, the number of heads is no more than 25.

4.3.14. *Answer*: 4 guineas.

Solution. Example. Alice payed with 3 coins of 13 guineas each, and she received 5 coins of 7 guineas each as the change.

Estimate. The face value of each coin is congruent to 1 modulo 6, so if Alice paid n coins and got $n + 2$ coins back, the purchase price is congruent to $n - (n + 2) = -2 = 4$ modulo 6. So Alice paid at least 4 guineas.

4.3.15. *Answer*: not possible.

Solution. Let R be the radius of the circle. Consider the boundaries of pieces that are arcs of radius R. Such boundaries can be either convex or concave. When reassembling, those boundaries that do not go along the boundary of the assembled figure lie inside of the figure adjacent to the boundary of "opposite convexity." Therefore, in the square, convex and concave arcs cancel out each other; that is, the difference in the sum of their lengths is zero. But for a circle, the difference between the sums of the lengths of convex and concave arcs is $2\pi R$. Therefore, the rearrangement of a circle into a square is impossible.

4.3.16. *Answer*: 6.

Solution. *Estimate.* Let $ABCD$ be the original quadrilateral F_1. We can assume that each time the half of the quadrilateral containing the side CD is flipped, the part containing the side AB remains fixed. Moreover, the sum of the angle A and the opposite angle does not change. In addition, the sides lengths do not change. But three values (i.e., lengths of sides other than AB) can be ordered only in 6 ways.

Let us prove that if two obtained quadrilaterals $ABKL$ and $ABMN$ have the lengths of the sides ordered identically ($BK = BM$, $KL = MN$, $LA = NA$), then the quadrilaterals are equal. It is enough to prove the equality of the diagonals AK and AM. Assuming that the diagonals are not equal, so that, say $AK > AM$, we get either $\angle ABK > \angle ABM$ or $\angle ALK > \angle ANM$, which contradicts the equality of the sums $\angle ABK + \angle ALK$ and $\angle ABM + \angle ANM$.

Example. Six different quadrilaterals come from any quadrilateral for which all side lengths are different, the sums of the opposite angles are also different, and which, when transforming *remain convex and do not degenerate into a triangle*. The only problem is that quadrilaterals may become equal when the order of the sides is reversed. However, they are not equal since when turning one of them over, the sums of the opposite angles will not be equal. This condition is satisfied, for example, for quadrilaterals whose area *is greater than half of the product of the two longest sides*; this inequality obviously does not hold for nonconvex and degenerate quadrilaterals. In particular, this condition is satisfied in the example below.

A specific example is the rectangular trapezoid with bases 3 and 6 and altitude 4.

4. Colorings

In problems on checkered boards, invariants most often take the form of some coloring of the cells.

4.A. Tilings (1) *By A. Ya. Kanel-Belov*

4.4.1. Two opposite corner cells were cut from a chessboard. Prove that the remaining piece cannot be cut into "dominoes" (pieces of 2×1 cells).

4.4.2. One 1×1 square containing the center was removed from the 4×4 square. Can the remaining shape be cut into 1×3 rectangles?

4.4.3. Where do you need to cut one cell from the 5×5 square to be able to cut the remaining shape into 1×3 rectangles?

4.4.4. Is it possible to cut a 10×10 square into figures as below consisting of four cells of size 1×1?

4.4.5. From a set containing several 2×2 squares and several 1×4 rectangles a 10×10 square is assembled, using all the pieces of the set. Then the mosaic was scattered and one of the 1×4 rectangles was replaced with a 2×2 square. Will it be possible now to assemble the 10×10 square?

4.B. Tables (2)[5] *By D. A. Permyakov*

4.4.6. In the cells of a $2N \times 2N$ table pluses and minuses are written down in some way. In one move, you can change the sign in all cells of any "cross", that is, of the union of some row and column. Prove that in a few moves you can get a table with all cells containing only pluses.

4.4.7. All cells of the 8×8 chessboard contain pluses, except one noncorner cell, which contains a minus. One can simultaneously change signs in all cells in one row or in one column or in one diagonal (a diagonal is the line along which a chess bishop moves). Prove that a board where all cells contain pluses cannot be obtained by such moves.

4.4.8. In each cell of an $n \times n$ table there is a minus sign. In one move you can change signs in one figure of Z-tetramino (i.e., in a checkered figure that is obtained by a translation, rotation, or reflection from a combination of cells a1, b1, b2, c2 of a checkerboard). For which values of n can you get a table with all pluses?

4.4.9. Is it possible to put the numbers $0, 1, 2$ in cells of a 100×100 table so that every 3×4 rectangle would contain 3 zeros, 4 ones, and 5 twos?

In conclusion, we present several problems for "estimate + example".

---

[5]Problems in this section are from All-Union Olympiads.

4.4.10. What is the largest number of checkers kings that can be placed on an 8×8 checkerboard so that every king can capture at least one other king (according to the rules of chess)?[6]

4.4.11. What is the smallest number of chips that can be placed on the chessboard of the size
   (a) 8×8,
   (b) $n \times n$
so that every line passing through the center of an arbitrary square and parallel to either a side or a diagonal of the board would meet at least one chip?

4.4.12. What is the smallest number of L-shaped three-cell pieces that need to be placed on 8×8 board cells so that no additional such piece can be placed on the board without overlapping with the existing ones?

4.4.13. A square 7×7 cell board is given. One is asked to mark the centers of n cells so that no 4 marked points are vertices of a rectangle with sides parallel to the sides of the square. What is the largest n for which this is possible?

Suggestions, solutions, and answers

4.4.1. Each "domino piece" contains one white and one black cell. But in our chessboard, there are 32 black and 30 white cells (or vice versa). For a detailed discussion, see [**Soi**].

5. Semi-invariants[7] (1)
By A. V. Shapovalov

If the word "invariant" means "invariable", then "semi-invariant" means half-invariable.[8]

It happens that when we change an object, some quantity associated with this objectn can only change in one direction, that is, either only increase or only decrease. It is also possible that as we make moves in one direction, a quantity associated with the position also changes in one direction. For example, when playing tic-tac-toe, the number of filled cells

---

[6]Excluding the "flying king" rule.

[7]The ideological predecessors of the "Invariants" and "Semi-invariants" sections are, among other things, the corresponding sections of the book [**K-BK08**].

[8]Another term used for semi-invariant is "monovariant", which emphasizes that the quantity, if it changes, does so *monotonically*, i.e., in only one direction.

with each move increases. Therefore, on a finite board, the game will end sooner or later. When playing on an endless board, the game may never end, but we can guarantee that the position does not happen again because the number of filled cells increases with each move!

A little more formally, this can be expressed as follows. Assume that we change an object (or a position) using *permitted operations* (or *moves*) and assume that with each object/position we can associate a *quantity* in such a way that with every permitted operation the value of this quantity either does not change or changes in one particular direction. Then this quantity is called a *semi-invariant*.[9] If the value of a semi-invariant changes with each operation/move, it is called *strict*; otherwise it is *nonstrict*.

In typical problems "on semi-invariant" they prove the impossibility of
(a) repetition of a position;
(b) an infinite number of moves;
(c) reaching certain positioins.

For the latter, one finds a semi-invariant and verifies that to obtain the desired construction from the original one, the semi-invariant should *change the wrong way*.

But how do we find a semi-invariant? Start with checking sample values: sums, products, areas, perimeters, and their combinations. If the object or position depends on integers, then the semi-invariant can be the GCD or LCM of these integers.

In the following two problems, it is important that the semi-invariant is an integer and that it cannot be greater than a certain number.

4.5.1. On a 100×100 chessboard, the king starts in the lower left cell and moves to the right, up, or right-up diagonally. What is the largest number of moves the king can make?

4.5.2. Integers are placed in cells of a 99×99 table. If in some row (or column) the sum is negative, one can change the signs of all numbers in this row (or column) to the opposite sign. Prove that only a finite number of such operations can be done regardless of what the original table was.

If the semi-invariant is not an integer, then its boundedness does not yet guarantee the end of the process (e.g., a decreasing positive semi-invariant can take infinitely many values $1, 1/2, 1/3, 1/4, \ldots, 1/n, \ldots$). In these cases, termination of the process is guaranteed by finiteness of the number of positions.

---

[9]This sentence is not a formal definition of a semi-invariant. But for the solution of problems a formal definition is not needed.

4.5.3. Suppose we are given 10 numbers. At one step, two nonequal numbers can be replaced by two equal number with the same sum. Does there exist an initial set of numbers for which the process
    (a) continues infinitely long;
    (b) loops (that is, can the same set of numbers occur twice)?

4.5.4. Several numbers are written on a circle. If for some four consecutive numbers a, b, c, d we have $(a - d)(b - c) < 0$, then the numbers b and c are swapped. Prove that this operation can be performed only a finite number of times.

Very often a position with no permitted operations is the desired one.

4.5.5. Numbers are written in the cells of a rectangular table. One can simultaneously change the signs of all numbers in a column or in a row. Prove that by repeating this operations, you can turn this table into one where the sum of numbers in any row or in any column is nonnegative.

In combinatorial problems, a semi-invariant often is the number of combinations, for example, pairs, triples, subsets, or permutations of some kind.

4.5.6. In the Far Far Away kingdom, all cities raised flags over the town halls, either blue or orange. Every day, residents recognize the colors of the flags of their neighbors within a radius of 100 km. One of the cities where the majority of neighbors have flags of a color different from the color of their flag changes the color of its flag to this different color. Prove that with time the color changes of the flags will stop.

Some constructions are created by "sequential improvements". We take the imperfect design and start to convert it. A semi-invariant guarantees that the process terminates and the desired effect is achieved.

4.5.7. In a parliament, each deputy has at most three enemies. Prove that parliament can be divided into two houses so that each deputy will have at most one enemy in his house.

4.5.8. On the plane, 100 red and 100 blue points are placed, no three on one line. Prove that one can draw 100 disjoint segments connecting these with endpoints of different colors.

A semi-invariant can also be *nonstrict*, i.e., it doesn't necessarily change with each move. Then it is helpful to find another semi-invariant which necessarily changes just when the first one does not change.

4.5.9. On a 100×100 chessboard a king is allowed to move to the right, up, right-up, or right-down diagonally. Prove that it can make only a finite number of moves.

If the second semi-invariant is not strict, then consider the third, the fourth, etc. In this case, it is natural to consider the sets of values of semi-invariants as strings ordered *lexicographically* (as words in a dictionary: the first elements are compared and if they are equal, then compare the second elements, etc., until the first mismatch).

4.5.10. In a deck, some of the cards are face up. From time to time Pete takes out from the deck a pack of several consecutive cards where the top card and the bottom card are face up (in particular, a pack can consist of just one face-up card), flips this pack as a whole, and inserts into the same place in the deck. Prove that no matter how Pete chooses the packs, eventually all cards will be face down.

In conclusion we present a few more problems on semi-invariants and their combinations.

4.5.11. There are several (a finite number of) numbers on a line. Every second, a robot picks a pair of adjacent numbers in which the left number is greater than the right, swaps them, and multiplies both numbers by 2. Prove that after several steps the robot will not be able to find an appropriate pair.

4.5.12. Carlson has 1000 cans of jam. Cans are not necessary identical, but each one contains not more than a hundredth part of all the jam. For breakfast, Carlson can eat the same amount of jam from any 100 of his cans. Prove that Carlson can act so that after some number of breakfasts he eats all the jam.

4.5.13. There are several positive numbers on a circle, each not exceeding 1. Prove that the circle can be divided into three arcs so that the sums of numbers on adjacent arcs differ by at most 1. (If there are no numbers on the arc, then the sum on it is considered to be zero.)

4.5.14. Ten people are sitting at a round table. In front of each person there are several nuts. The total number of nuts is one hundred. At a signal, each one transmits part of her nuts to the neighbor on the right: half if the giver has an even number of nuts, or one nut plus half of the remainder if she has an odd number of nuts. Such an operation is performed a second time, then a third, and so on. Prove that after some time everybody will have exactly ten nuts.

Suggestions, solutions, and answers

4.5.1. *Answer*: 198.

Solution. We enumerate the columns from left to right, and the rows from bottom to top. By the weight of a cell we mean the sum of the numbers of the row and the column containing this cell. Let us look at the weight of the cell the king stands on. This is a semi-invariant; with the allowed move, it increases by 1 or by 2. The smallest possible weight is $1 + 1 = 2$, and the largest is $100 + 100 = 200$. The increase in weight does not exceed $200 - 2 = 198$, which means that the number of moves is also no more than 198. Such a number of moves is possible if the king passes from the lower-left cell to the upper-right without making moves diagonally (for example, first 99 moves to the right, and then 99 moves up).

4.5.2. *Solution.* If there was a negative sum $-A$, then by the change of signs in this series the sum becomes equal to A. Therefore the sum S of all table elements will increase by $2A$. So S is a semi-invariant. Under permitted operations, the absolute values of elements do not change; that is, the sum T of all absolute values is an invariant. Moreover, $S \leq T$ always. Since A is a natural number, $A \geq 1$ and each time S increases by at least 2. Therefore, the number of operations is finite (the maximum increase is from $-T$ to T with step 2, so there are no more than $2T/2 = T$ operations).

4.5.3. (a)*Answer*: it can.

Solution. Suppose initially there are the following 10 numbers: 2, 4, 4, and seven zeros. We will perform operations without touching zeros. At the first step, replace 2 and 4 with 3 and 3. We get the set consisting of 3, 3, 4, and zeros. Further, having positive numbers a, a, b (where $a \neq b$), we replace a and b. We get the three positive numbers a, $(a + b)/2$, $(a + b)/2$. Note that $(a + b)/2 \neq a$ (equality is possible only for $a = b$). We are again in a situation where two numbers are equal and the third is different from them. Continuing such actions, we will never make all numbers equal.

(b) *Answer*: cannot.

Solution. The process has a strict semi-invariant: the sum Q of squares of all numbers. Indeed, we can assume that with each step we replace the numbers $a + x$ and $a - x$ (where $x \geq 0$) with a and a. The old numbers contributed $2a^2 + 2x^2$ to Q, and the new ones contribute $2a^2$, so Q decreased. Therefore, repeating the set is impossible.

Remarks. 1. Other semi-invariants are also possible, for example, the sum of pairwise products. However, the product of all numbers cannot always be the semi-invariant.

2. The semi-invariant Q decreases but cannot become negative. It would seem that this contradicts the possibility of an infinite number of operations. But it only means that the changes in Q can be getting smaller and smaller.

4.5.4. *Solution.* Consider the sum Q of squared differences of neighboring numbers. During the operation, only two terms will change; instead of $(a-b)^2 + (c-d)^2$ we get $(a-c)^2 + (b-d)^2$. As a result, the difference between these sums, i.e., the number $2(a-d)(b-c)$, will be added to Q. By the conditions, this addition is negative; therefore, Q will decrease. The existence of the semi-invariant Q guarantees that at each step we get an arrangement not previously encountered. But the total number of rearrangements is finite (for n numbers on the circle it is no more than $(n-1)!$); therefore the number of operations is finite.

4.5.5. *Solution.* We will always change signs in the row or column where the sum of the numbers is negative. Then the sum of the numbers S in the entire table will increase. The semi-invariant S guarantees nonrepetition of the arrangement of signs of numbers of the table. But the number of all combinations of signs is finite (it equals 2^N, where N is the number of nonzero numbers in the table). So, at some point we cannot get a new arrangement and the next step is impossible. This means that the sums of numbers in all rows and columns has become nonnegative.

4.5.6. *Solution.* The number of pairs of neighboring cities with flags of different colors is a semi-invariant. This number decreases every day by at least 1, so the number of days with changing colors of flags is no more than the total number of pairs of cities.

4.5.7. *Solution.* First, we divide the deputies into two chambers arbitrarily. If some deputy finds at least two enemies in her chamber, we transfer her to another chamber. After this move the number of pairs of enemies in the old chamber decreases by at least 2, and the number of pairs of enemies in the new chamber increases by at most 1. Thus, the total number of pairs of enemies in the same chambers will decrease; i.e., this number is a semi-invariant. It decreases by at least 1, and at some point further decrease will be impossible. Therefore, the transition is also impossible; that is, no deputy has two or three enemies in her chamber.

4.5.8. *Solution.* Draw segments with endpoints of different color arbitrarily. Let S be the sum of the lengths of the segments. If a pair of intersecting segments AB and CD is replaced with the pair of segments AC and BD or with the pair AD and BC, then by the triangle inequality the sum of the lengths of the pair will decrease. For a pair of intersecting segments with endpoints of different color choose a pair of endpoints of the segments with different colors and replace the original pair of segments with the new pair. For such replacements, S is a strictly decreasing semi-invariant. The number of ways to draw segments with endpoints of different colors is finite (it is equal to 100!). The existence of a semi-invariant and the finiteness of the number of possible "configurations" guarantee that the process ends, and in the final position the segments do not intersect.

4.5.9. *First solution.* Similarly to the solution of problem 4.5.1, we enumerate the columns and the rows and define the weight of the cell as the sum of these numbers. This is a nonstrict semi-invariant: the weight does not change during right-down moves. Add the number of the column as the second semi-invariant. This number is a semi-invariant for those moves that do not change the first semi-invariant. Now it is clear that a series of consecutive moves right-down cannot be infinite (in fact, there are not more than 99 such moves). So, we have a finite number of weight increases and between two increases a finite number of moves. Therefore, the total number of moves is finite.

Second solution. Let a cell lie in the vth column and the hth row. To such a cell we associate the number $S = 2v + h$. Then S is a strict semi-invariant: for admissible moves S increases by 1, 2, or 3 (in particular, $2 - 1 = 1$ for right-down moves). So, since the number of cells is finite, the number of admissible moves is also finite.

Note. The second solution is shorter, but the first is more natural and easier to come up with. Almost always a set of semi-invariants can be replaced with one semi-invariant which, however, may look artificial.

4.5.10. The position of cards in the deck can be encoded with an n-digit number; going from top to bottom, we write 1 for the card that lies face down and 2 for the card that lies face up. For a permitted operation this number will decrease; to some place in the higher ranks nothing changes, then 2 changes to 1, and maybe there are more changes within lower digits. So the code number is semi-invariant, it decreases by at least 1 at each step, and it cannot become negative. Therefore sooner or later the operations will become impossible, which means that all cards lie face down.

Comment. Actually, comparing multiple-digit numbers is similar to comparing words in a dictionary. We can encode a deck not with numbers, but with the letters D and U (meaning "face down" or "face up"). Then, at each operation, the code word will be replaced by a word closer to the beginning of the dictionary; that is, the index of the word in the dictionary will be a semi-invariant. This encoding is more convenient when there are more letters than digits, and it is crucial when the code sequences are infinite.

4.5.11. *Suggestion.* Suppose the numbers are written on cards and the robot rearranges the cards and changes the numbers on them. Prove that no pair of cards can be swapped more than once.

Solution. See Problem 23.3.6 in the book [**MedSha**].

4.5.12. *Suggestion.* We say that a nonempty can is overfull, full, or incomplete if it contains more, equal to, or less than one hundredth of the jam remaining at this moment. There are no overfull cans at first. Show that Carlson can eat jam every day (except for the last) in such a way that overfull cans do not appear, but the number of incomplete ones decreases.

Solution. See Problem 27.3.6 in [**MedSha**].

4.5.13. Let the *weight* of an arc be the sum of the numbers on it (an arc without numbers has weight 0), and let the *sparseness* of an arc be the difference between the largest and the smallest numbers on this arc. The number of partitions with different weights is finite. We choose a partition with *minimal* sparseness and prove that this is the required partition. Suppose the difference between the largest weight c on the arc C and the smallest weight a on the arc A is greater than 1. We move the boundary between the arcs so that exactly one number r moves from C to A. After that, we get a new partition: into arcs A_1, B, and C_1 with the sums $a_1 = a+r$, b, $c_1 = c-r$. It is easy to verify that each of the differences $c_1 - a_1$, $b - a_1$, $c_1 - b$ is less than $c - a$ but greater than -1; i.e., the sparseness has decreased, which contradicts the choice of the initial partition.

4.5.14. *Suggestion.* Prove that

(a) the sum of the absolute values of the differences of each person's number of nuts and the number 10 is a nonstrict semi-invariant and

(b) if this semi-invariant does not change, then the number of consecutive zero differences between a positive and a negative difference increases.

Solution. See Problem 19.7.6 in [**MedSha**].

Chapter 5

Algorithms

1. Games (1)[1]
By D. A. Permyakov, M. B. Skopenkov, and A. V. Shapovalov

Using specific examples, we will get acquainted with some beautiful ideas of game theory. General guidelines on the topic "Games" can be found in the corresponding section of the book [**FGI**].

Symmetric strategy

The most common strategy in games is the *symmetric* strategy (and also its generalization, the *complementary* strategy). To solve the following problems it is useful to be familiar with Section 2 of Chapter 4 since many strategies in games are based on invariants (an example of an invariant is the position symmetry).

5.1.1. (a) Two people in turn put domino pieces on the 8×8 chessboard. Each domino covers exactly two cells of the board, and each cell can be covered with at most one domino. The player who can't put another piece loses. Who wins if both players make the best moves? How should they play?
(b) The same questions for the 8×9 board.

Let us explain what the questions mean. To answer the first question you need to name the player who wins with *any* moves of her adversary. To answer the second question you need to describe an *algorithm* of how to choose the moves of this player that guarantees that she wins (i.e., the *winning strategy*). It is important to clearly separate an algorithm *itself* from a *proof* that the algorithm leads to the desired result.

Part (b) of this problem shows that symmetry of position does not guarantee that the symmetric strategy works.

---

[1]Sections "Symmetric Strategy", "Growing tree of positions", "Passing the move" were written by D. A. Permyakov and M. B. Skopenkov; "Joke games", "Games on outracing", "Accumulation of advantages" were written by A. V. Shapovalov; "Miscellany" was written by all three authors.

5.1.2.° (Challenge) What will the result be if Black attempts to mirror (i.e., symmetrically copy) the other player's moves in regular chess in the case where White plays in the optimal way? Choose the correct answer:

(1) draw;   (2) White wins;   (3) Black wins.

The key idea is not so much symmetry but breaking up all possible positions in pairs. *Complementary strategy* is to respond to the opponent's move by a move to the second position of the corresponding pair.

5.1.3. A king is standing on a chessboard. Two players take turns moving it. A player loses if after her move the king ends up in a cell it visited earlier. Who wins if both players play optimally?

Game on outracing

Game on outracing is a common trick in nonmathematical games. But in math games it happens that the gain goes to the first one who will be able to occupy a key position. After that, as a rule, a complementary strategy works.

5.1.4. There are 9 sealed transparent boxes, respectively, with $1, 2, 3, \ldots, 9$ chips. Two players take turns taking one chip from any box, unsealing the box if necessary. The last player who is forced to remove the seal from a box loses. Who will win regardless of the opponent's moves?

5.1.5. In one of the corners of a chessboard there is a flat cardboard 2×2 square, and in the opposite corner there is a 1×1 square. Two players take turns rolling squares across a side: Bob rolls the large square, and Mike rolls the small one. Bob wins if no later than on the 100th move he moves to cover the cell where Mike's square is. Can Bob win regardless of Mike's game if

(a) Bob starts first;

(b) Mike starts first?

Accumulation of advantages

Accumulation of advantages is also a very common trick in nonmathematical games. In math games, accumulation is usually associated with some kind of semi-invariant. Therefore to study such games acquaintance with Section 5 of Chapter 4 is useful. Here you need to find an algorithm leading to accumulation of advantages regardless of the opponent's resistance.

5.1.6. Mike is standing in the center of a round lawn of radius 100 meters. Every minute he takes a step one meter long. Before every step he announces the direction in which he wants to move. Kate has the right to make him change to the opposite direction. Can Mike act in such a way that at some moment he will leave the lawn, or can Kate always prevent this?

5.1.7. On a $1 \times 100\,000$ grid strip (initially empty), two players make moves in turns. The first one can put two checkmarks on any two free strip cells. The second can erase any amount of checkmarks in a row (without empty cells between them). If after the move of the first player there are 13 or more checkmarks in a row, she wins. Can the first player win if both players play optimallly?

5.1.8. Two players take turns breaking a stick: the first player breaks it in two parts, then the second player breaks any of the pieces into two parts, then the first breaks any of the pieces into two parts, etc. A player wins if after one of his moves he can build two congruent triangles using six existing pieces. Can the other player prevent this from happening?

Joke games

In *joke games* one of the players always wins regardless of whether she wants to or not.

5.1.9. (a) On the table there 2015 piles of one nut each. In one move, you can combine two piles into one. Two players make moves in turn, and the one who cannot make a move loses. Who wins?

(b) The same question when one can combine piles with only the same number of nuts.

5.1.10. Given the $1 \times N$ grid strip, two players play the following game. On each turn, the first player puts a cross mark in one of the free cells, and the second one puts a zero. One cannot put two crosses or two zeros in adjacent cells. The player who cannot make a move loses. Which of the players wins if both make optimal moves? What is the winning strategy?

In addition to joke games, there are *almost joke games*, where the winning strategy is as follows: if there is a win in one move, this move must be made, and otherwise, you can make any move. Another version allows a player to make any move except those that lead to an immediate loss for this player. In such games, it is important to guess who will have the opportunity to make a winning move or who will be forced to make a losing move and to prove this. In addition, a winning strategy may be to get to a position after which the game turns into a joke game with the desired outcome.

5.1.11. Ten baskets contain 1, 3, 5, ..., 19 apples, respectively. First Bill takes one apple from any basket, then Gene takes one, then Leo does, then

Bill again, etc. A player loses if after his move there will be an equal number of apples in some of the baskets. Which player cannot avoid losing?

5.1.12. A checkered 9×9 square is made of matches; each side of each cell is one match. Pete and Bill take turns removing the matches one at a time, with Pete starting. A player wins if after his move there are no 1×1 squares anymore. Which player can win regardless of how his opponent plays?

Growing a tree of positions

One of the universal ways to analyze a game is *growing a tree of positions*.

5.1.13. *Put the Queen in the Corner, or "tsyanshidzi".* The queen stands in square d1 of the chessboard. Two players take turns moving it up, right, or right-up. The one who puts it in the square h8 wins. Who will win if both players play optimally and how should the winning player play?

If you cannot solve it, think about the next question first.

5.1.14.° Who wins the game in the previous problem if initially the queen stands in the square f4? Choose the correct answer:
(1) the first player;   (2) the second player.

Growing a tree of positions means a complete analysis of the game. Let us go now to the more complicated idea of *passing the move*, which helps even when there is no possibility for a complete analysis.

Passing the move

5.1.15. In *two-move* chess, pieces move according to the usual rules, but with each turn a player can make exactly two usual moves with the same piece. The goal of the game is to checkmate the opponent's king. The rules of repeating the position three times and of 50 moves do not apply.[2] Prove that White can play two-move chess so that they will not lose (i.e., they either win or draw).

5.1.16.° Rules of *chess without zugzwang*[3] differ from the rules of ordinary chess just by adding the opportunity to skip the turn. Can Black win if White plays optimally? Choose the correct answer:
(1) they can;
(2) they cannot.

---

[2] If you don't know what the rules are, ignore this sentence.

[3] *Zugzwang* in chess is a position for a player in which any move makes it worse.

Miscellany

In the following problems it is necessary to find which player wins if both players make the best moves. If the statement of the problem depends on parameters m and n, then the answer may also depend on m and n.

5.1.17. Black and white rooks stand in adjacent corners of the 9×9 chessboard, and the remaining cells are occupied by gray pawns. Two players take turns each with their rook, and with each move they need to capture either a gray pawn or a rook of the opponent. The player who cannot make a move loses.

5.1.18. On one table there are 34 stones, and on another one there are 42. Two people play the following game. At each move the number of stones equal to a divisor of the number n (possibly all n stones) can be taken from the table with n stones and placed on the other table. A player loses if after her move the pair (a, b) of the number of stones on the tables coincides with one already encountered in the game. Pairs (a, b) and (b, a), $a \neq b$, are considered as different.

5.1.19. On an endless grid paper, two players in turn paint over the segments between adjacent grid nodes, each with their own color. The goal of the first player is to get a closed broken line of her color. There is no upper bound on the lengths of the game. Can the second player stop his opponent?

5.1.20. The number 2 is written on the blackboard. At each move one can add to the number on the board any of its divisors that is smaller than the number itself. The winner is the first player to get a number greater than $1\,000\,000$.

5.1.21. On the table there are two piles of matches: m matches in one pile, n matches in another one, $m > 2n$. Two people are playing the game. At each move a player can take from one of the piles a nonzero number of matches that divides the number of matches in the second pile. The player taking the last match from one of the piles wins.

5.1.22. The city is a rectangular 10×12 grid of streets. Two companies take turns placing lanterns at unlit intersections. Each lantern illuminates a rectangle in the city with a vertex in this lantern and with another vertex in the north-east corner of the city. The company that lights up the last intersection loses.

5.1.23. Two players play the following game on an $m \times n$ board. They have a white and a black king, respectively, standing in the opposite corners of the board. They move their kings (according to the rules of chess) alternately so that the distance between the centers of the cells in which kings stand decreases (kings are allowed to occupy neighboring cells, but they can't stand in the same cell). The player that cannot make a move loses.

5.1.24. Given an empty table of size 13×17, two players take turns putting chips in empty cells. The first player can put the chip at the intersection of a row and a column if the row and the column together contain an even number of chips, and the second player can put a chip if the combined number of chips is odd. The player who cannot make a move loses.

5.1.25. Let two piles of matches be given, one with 1997 matches, another with 1998 matches. Two people play the following game: each player throws away one of two piles and divides the other into two nonempty piles of not necessarily equal size. The player who cannot split a pile in two parts loses.

5.1.26. A positive integer is called *permitted* if it has at most 20 different prime divisors. At the beginning there is a pile of 2004! stones. Two players take turns removing a permitted number of stones from the pile. The last player able to play wins.

Several other interesting games are analyzed in articles [**Pev, Gik, S72**].

Other common techniques used in the elementary theory of games (but which we don't consider in this book) are "greedy algorithm" and "menagerie".

In all the games we reviewed, the position was completely known to all players. These are the so-called *games with perfect information*. For practical applications, games with incomplete information are no less important; you can read about them in [**Tm, Ver, ChS**].

Suggestions, solutions, and answers

5.1.1. (a) *Answer*: the second player wins. The algorithm is as follows: for each move of the first player the second player responds with a move symmetrical about the center of the board.

Solution. We prove that the second player can always make the move prescribed by this algorithm without violating the rules.

Denote by n the number of moves the first player made from the beginning of the game. We prove by induction on n the following stronger statement: *if the first player was able to make the nth move, then the second player will be able to respond with a move symmetric about the center of the board, and the position will remain centrally symmetric.*

Induction base: $n = 0$. The starting position is centrally symmetric. The first player has not yet made any move, so nothing should be proved.

Induction step. After the nth move of the second player a centrally symmetric position occurred and the first player put dominoes on the cells A and B. The second player responds with a move to the cells A' and B' symmetric to them about the center of the board. Let us prove that the cells A' and B' are empty. They are not affected by the last domino of the first player, since $A' \neq A, B$ and $B' \neq A, B$, because on the 8×8 board two centrally symmetric cells cannot coincide nor can they be adjacent on a side. Previous dominoes on the board cannot cover A' and B' either, since after the previous move of the second player the position was centrally symmetric. Therefore, the described $(n + 1)$st move of the second player is possible. It is clear that the position remains centrally symmetric after making it, and the required statement is proved.

It is clear that the game on the finite board cannot go on indefinitely, so in some moment the first player cannot make a move and loses.

(b) *Answer*: the first player wins using the following algorithm. The first move is to put a domino at the center of the board (i.e., on two cells, adjacent to the center) and then to copy the moves of the second player symmetrically about the center.

5.1.3. *Suggestion*: *an alternative strategy.* Let us prove that the first player wins. We break the board into "dominoes", i.e., pairs of cells adjacent side by side. At each move, the first player moves the king to the second cell of the domino where he was before the move. After each move of the first player, the king will visit either two or zero cells of each domino. Therefore, after the move of the second player, the first player can make the move according to the described strategy.

5.1.4. *Answer*: the second player wins.

Suggestion. The second player should make sure that after their fourth turn only boxes with an odd number of chips remain sealed. Since there are more odd boxes, at least one box with an odd number of chips will not be opened. Therefore, such a box will be unsealed last. But the total number of chips in the remaining boxes is even, so all boxes can become empty only after a move of the second player; i.e., the first player will be forced to unseal the last box.

5.1.5. (a), (b) *Answer*: no.

Suggestion. Let the 2×2 square cover cells $a1$ and $b2$ and the 1×1 square cover the cell $h8$. Let us call the upper-right vertex of the cell $f6$ node A and the upper-right vertex of the cell $d6$ node B. Mike should move his square to the corresponding node (A in case (a), to B in case (b)) and run around it avoiding covering.

5.1.6. Mike should increase the distance from the center of the lawn while moving perpendicularly to the radius passing through the point at which he stands at the moment.

5.1.7. *Answer*: yes, he can.

Suggestion. The first player starts with increasing the number of groups of 13 squares with six "holes", then increasing the number of groups with five holes, etc.

5.1.8. *Answer*: no.

Suggestion. The first player breaks the stick in half and then repeats the moves of the second player until 5 pairs of equal pieces of lengths $a \geq b \geq c \geq d \geq e$ are formed. If a triangle cannot be formed from any triple, then the first breaks off a piece of length c from a. The second player must break c (otherwise the first one will break off a piece of length c from b and can build two isosceles triangles either with sides c, c, d or with sides c, c, e). The first player breaks off a piece of length c from the other piece of length a and then breaks one by one pieces of length c each from $a - c$, etc., until it becomes possible to build two triangles from pieces of length $a - kc$, b, and c.

5.1.9. In both cases, the total number of moves does not depend on the course of the game. In part (b), the end position is determined by the binary expansion of the number 2015.

5.1.10. After the return move to one of the endpoints of the strip, the second will win regardless of the moves of the first player .

5.1.11. *Answer*: Gene.

Solution. In all baskets there are 100 apples total. Note that there is exactly one *hopeless* position, where *all moves* are losing; this is the position where there are $0, 1, 2, \ldots, 9$ apples in the baskets. In any other position one can avoid losing in one move by taking an apple from the smallest basket if it contains at least one apple, or from a basket with the number of apples different from the previous one by 2. In the hopeless position there are 45 apples, so, before getting into it 55 apples should be taken. If no one loses before, then Bill will take the 55th apple and then Gene should take the 56th apple. Thus, only Gene can get into a hopeless position. But then Leo and Bill always have the possibility of not losing in one move. Let them do such nonlosing moves. Since the game lasts at most $1 + 3 + \cdots + 19 = 100$ moves, it will eventually end, and therefore, Gene will lose. (He will either lose before he takes the 56th apple or when he falls into the hopeless position.)

5.1.12. *Answer*: Bill will win.

Suggestion. After Pete's move there always remains an odd number of matches for Bill. If there are two complete squares without a common side, they use 8 matches. Bill takes any match not from these squares (because there is one), and at the next move Pete cannot destroy both squares. If there are not two such squares, then there is a single complete square or two complete squares with a common side. In both cases Bill will be able to destroy them and win.

5.1.13. *Answer*: the second player wins. The algorithm is as follows: after each move of the first player, the second player should place the queen at one of the squares e3, f7, g6, h8. (Strictly speaking, for a complete description of the algorithm, for each position of the queen one should indicate at which of these cells the queen needs to be moved. We leave this to the reader.)

Solution. Considering all possible moves of the first player starting with squares d1, e3, f7, g6, we show that the return move to one of these cells or to h8 is always possible.

Path to solution. Our goal is to put in each square of the 8×8 board the sign "+" or the sign "−" depending on whether or not the next move can win when the initial position of the queen is at this cell. The positions with the "+" sign are called *winning*; the remaining ones are called *losing*. We will fill chessboard squares sequentially starting with the 8th horizontal, the column "h", and the diagonal "a1–h8". It is clear that in any square of the 8th horizontal, the column "h", and the diagonal "a1–h8" (except the square h8) you need to put the sign "+". It is easy to see that the remaining squares are filled one by one using the following two "golden" rules:

(1) if a permissible move from a square can be made to a square with the sign "−", then this square should get the sign "+";

(2) if after each permissible move from a given square we end up at a square with the sign "+", then this square should get the sign "−".

Using these rules, we will fill all the squares of the chessboard. The squares h8, g6, f7, e3, d1, c5, a4 will get the sign "−", and the remaining ones will get "+". Once the square d1 gets the sign "−", the second player wins. His winning strategy is "Put it on a square with the minus sign!"; i.e., after each move of the first player, he must again place the queen in the square with the "−" sign (i.e., in one of the squares c3, f7, g6, h8).

Comment. Details of this problem and its generalization to the board of arbitrary size in which Fibonacci numbers arise are discussed in articles [**Or77, Yag71, MS**].

5.1.15. Suppose that Black has a winning strategy. Then White makes the first move b1-c3-b1 (two consecutive knight jumps after which it returns to the original square) and then plays using this winning strategy for Black. Since this strategy is winning, White will win. The resulting contradiction shows that White *has a nonlosing strategy*. The details of the proof are discussed in the book [**KG**].

5.1.17. The second player wins, responding to each move symmetrically with respect to the vertical axis of symmetry.

5.1.18. *Answer*: the second player wins.

Suggestion. We will denote the position by the number of stones on the first table. Let the second player break all possible but not yet met positions into pairs that differ by 1: $(0, 1)$, $(2, 3)$, ..., $(32, 33)$, $(35, 36)$, ..., $(75, 76)$. On any move of the first player the second responds by shifting one stone to get the second position from the same pair. The second always has a move, so he will not lose, but since the game is finite, he will win.

5.1.19. *Answer*: yes.

Suggestion. The second player unites in pairs the segments bounding each cell from top and left. He responds to any move of the first player with the move in the same pair. Then on any colored closed polygonal line the highest among the leftmost vertices will be the endpoint of two distinctly colored segments.

5.1.20. *Answer*: the first player wins.

Suggestion. We prove this with the help of passing the move. After the first and second moves, the number 4 is obtained. If, starting with the number 6, the first player wins, then he now adds 1, receiving 6 after the response of the second player. If, starting with the number 6, he must lose, then he adds to 4 its divisor 2, and then the second player loses.

5.1.21. *Answer*: the first player wins.

Suggestion. We prove this with the help of passing the move. Let k and r be the quotient and remainder on division of m by n. Note that $k > 1$. Therefore, on his move the first player can get either kn matches or $(k-1)n$ matches from the first pile, getting either the position (r, n) or the position $(n+r, n)$. Note that from position $(n+r, n)$ there exists only one move and it leads to position (r, n). If the position (r, n) is losing for the next player, then the first one goes to it and wins. Otherwise, he goes to the position $(n + r, n)$; in response he gets (r, n) and wins.

5.1.22. *Answer*: the first company wins.

Suggestion. We prove this with the help of passing the move. If the second company has a winning strategy, then, in particular, the second company can win if the first move of the first company is to the north-eastern vertex. But then the first company can start using this strategy from the beginning!

5.1.23. *Answer*: let $m \leq n$. The second player wins if $m = 1$ and n is even or if m is an odd number greater than 1 and n is odd; otherwise the first player wins.

Strategy. Let x and y be the absolute values of the difference between row numbers and column numbers in which the kings stand; for definiteness $x \leq y$. We call a pair (x, y) *bad* if x is even and either $x = 0$ and y is even or $x > 0$ and y is odd; otherwise the pair is called *good*. Note that the only pair from which there is no move is a bad pair $(0, 1)$. With the rectangle sizes listed as winning for the first player in the answer above, the initial pair of differences is good. A winning strategy: make a move by decreasing x, y, or both and turning a good pair of differences into a bad one (this is always possible).

Since the bad pairs differ in some coordinate by at least 2, a move from a bad pair to a bad pair is impossible (even if one could increase the differences).

Path to solution: *growing a tree of positions*. Put in the cell with coordinates (x, y) of the $m \times n$ table a "+" sign if when playing on an $x \times y$ board, the first player wins, and put a "−" sign if the second player wins. We will fill in the table sequentially starting from cells $(1, 2)$ and $(2, 1)$. In these cells put a "−" sign. To expand the table we will use the following rules:

(1) if a king can move from a given cell to a cell in which the sign "−" stands and so that the distance to the center of the cell $(1, 1)$ decreased, then put in this cell the sign "+";

(2) if, after each possible move of the king from the given cell (such that the distance to the center of the cell $(1, 1)$ decreases) we get into the cell with a "+" sign, then put in the given cell a "−" sign.

Comment. You need to pay attention to the fact that moves are possible in which the distance between the centers of the cells on which the kings stand decreases, but the difference in the coordinates of these centers along one of the axes increases.

5.1.24. *Answer*: the first player wins.

Suggestion. Before the first player's next move, the number of empty cells is odd. We call a line (row or column) "even" if it has an even number of empty cells; otherwise it is called "odd". If the first player has no moves, then each empty cell lies at the intersection of an even and an odd line. In every even column, the number of such cells is even, and in every even row, it is also even; hence, the total number of empty cells is even. This is a contradiction.

5.1.25. *Answer*: the first player wins.

Suggestion. The first player wins, leaving two odd heaps after his move. In the return move, if the second one can make a move, the first will receive even and odd heaps. Dividing the even, he again leaves two odd heaps.

5.1.26. *Answer*: the second player wins.

Suggestion. Consider the least nonpermitted number P, which is obviously the product of the first 21 smallest primes. Each factor is less than 2004; therefore 2004! is a multiple of P. Multiples of P have at least 21 prime factors; therefore they are not permitted. The strategy of the second player: take the number of stones equal to the remainder on division by P. Then he will leave for the first player a number of stones that is a multiple of P. Taking away the permitted number that is not a multiple of P, the first player will leave for the second one a number of stones that is not a multiple of P. Therefore, the first player will not be able to take the last stone and thus will lose.

2. Information problems (2)
By A. Ya. Kanel-Belov

... in what year did the doorman's grandmother die?

What prevents us from answering the question in the epigraph? Not enough information!

Consideration of the amount of information also helps in mathematical problems. Often the necessary knowledge is not obtained immediately, but it is necessary to move to it in small steps. Each step imposes some restrictions on the final result, that is, adds some information. Sometimes the limitations obtained in several steps are not enough to fix a single result; this means that the result cannot be achieved with such steps.

5.2.1. Bill chose a number from the first hundred integers, and Nick is trying to guess it by asking questions. Bill answers all questions only with "yes" or "no". What is the minimum number of questions Nick needs to ask to determine the chosen number?

If we only know that the result has been arrived at, but the result itself is not yet known, sometimes we can determine the result.

5.2.2. (a) The king reported two consecutive positive integers to two of his sages. Neither of the sages heard what was told to the other, but both know that their numbers are consecutive natural numbers. They in turn ask each other a question: "Do you know my number?" Prove that at some moment the answer will be "Yes, I do."

(b) The same situation, but one was informed of the sum of two natural numbers, both less than 1000 (not necessarily consecutive), and the second was shown their product.

5.2.3. Two friends have not seen each other for many years. When they met, they started talking, and one boasted to the other that he already has three children. "How old are they?", asked the second. "The product of their ages is equal to 36, and the sum is equal to the number of this tram." Having a look at tram number, the second interlocutor said that this data was not enough. "And my eldest son is a redhead." "Then I know how old they are", said his friend and accurately gave the age of every child. How old was each child?

5.2.4. There are two cities in Zurbagania. In one of them live knights who only tell the truth, and in the other live liars who lie all the time. Regardless of their notions of honesty, residents of each city sometimes go to the other city for a visit. A traveler was in one of the cities and met a passerby. How can the traveler find out from the minimum number of questions who is in front of him and in what city he is located?

In all the problems considered above, there was a *transfer* of information. How do we measure the *quantity* of the transmitted information? By how much has this information helped to narrow down the list of possible options of the recipient? The fewer options left after the next step, the more information. Therefore in many informational problems one counts the number of options.

5.2.5. (a) A gold chain consists of 23 links, each of which costs one dollar. Which two links need to be sawn so that you can pay for any purchase worth a whole number of dollars between 1 and 23 without getting any change back? (The sawn link also costs one dollar.)

(b) Prove that if there are more than 23 links in a chain, then it cannot be sawn as required in part (a).

(c) What is the greatest length a gold chain can have such that after cutting three links in it, you can pay for any purchase worth from one dollar to the price of the entire chain?

5.2.6. Some of the 20 identical looking metal cubes are aluminum; the rest are duralumin, which is heavier. Cubes from the same material weigh the same. How, using a balance beam without weights, in no more than 11 weighings can one determine the number of duralumin cubes? The balance beam scales only show if the weights in the two cups are equal, and if they are not equal, they show which cup is heavier.

5.2.7. Of 81 identical looking coins, one is fake. What is the minimum number of weighings on a balance scale without weights required to find the counterfeit coin if it is known that it is lighter than the real ones?

5.2.8.* Of 12 coins, one is fake, and it is not known whether it is lighter or heavier than the others. What is the minimum number of weighings with a balance beam without weights required to find this coin and to determine whether it is lighter or heavier than the others?

5.2.9. There is a dot in a square of side 1. What is the minimum number of questions required to determine the coordinates of the dot up to 0.1?

5.2.10. Radioactive balls. Among n balls, two are radioactive. A measuring device indicates whether there are radioactive balls in the tested group. What is the maximal value of n for which one can determine at least one of the radioactive balls in k measurements?

Suggestions, solutions, and answers

5.2.1. *Answer*: 7. See article [**Or76**] where similar problems are considered.

5.2.2. (a) Consider how many questions will be asked if the second sage was told the number is 1? And if he was told the number is 2?

5.2.5. (a) *Answer*: for example, 4th and 11th.

5.2.7. For the sake of simplicity, you can first analyze cases of 3, 9, and 27 coins.

5.2.8. This problem and its generalization for any number of coins are discussed in detail in the article [**Shes**].

5.2.10. This problem is discussed in detail in [**Vil71a**, Solution of Problem M28].

3. Error correction codes (2)
By M. B. Skopenkov

In practice, not all information we receive is reliable. Therefore methods were invented to correct possible errors. The main result of this section is problem 5.3.3.

5.3.1. To transmit messages electronically, each letter of the English alphabet is represented in the form of a five-digit combination of zeros and ones corresponding to the binary notation of the number of a given letter in the alphabet (numbering of letters starts from one). For example, the letter "A" appears in the form 00001, the letter "B" becomes 00010, the letter "H" becomes 01000, etc. The five-digit combination produced is transmitted using a cable containing five wires. Every binary digit is transmitted by a separate wire. When receiving a message Crypto crossed the wires and instead of the word transmitted it received a set of letters "EREPNRX". Find the word passed.

5.3.2. Twenty-six letters of the English alphabet are encoded by sequences of zeros and ones.

(a) If, upon receipt of a message, an error in at most one bit is possible, then for the unambiguous recovery of the message, codes of different letters must differ in at least three digits.

(b) If messages consist only of letters A, B, C, D, E, F and when receiving a message an error in at most one bit is possible, then 5 bits are not sufficient for encoding.

(c) If, upon receipt of a message, an error in at most one bit is possible, then 8 bits are not sufficient.

(d) If an error in at most two digits is possible, then 10 digits are not sufficient.

(e)* Find the minimal number of bits sufficient to encode a message in (c).

5.3.3. Pete selected an integer between 1 and 2016. Bill can ask questions of the form "Does your number belong to this set?" Prove that Bill will be able to correctly guess the selected number in 15 questions assuming that Pete is allowed to lie once.

This is not an easy problem. To solve it, it is convenient to set the coding method in the "chess language". The subsequent "chess" problems lead to a solution.

5.3.4. What is the maximum number of rooks you can put on a chessboard so that each rook threatens at most two others? (Rooks cannot threaten *through* one another.)

5.3.5. (a) Given a $4 \times 4 \times 4$ cube, arrange 16 rooks in it so that they do not threaten each other.

(b) What is the maximum number of rooks that can be placed in an $8 \times 8 \times 8$ cube so that they do not threaten each other?

5.3.6. (a) Create 16 sequences of length 9 consisting of 0 and 1, any two of which differ in at least three digits.

(b) The same question, about 2^{n^2} sequences of length $(n+1)^2$.

(c) The same question, about 16 sequences of length 7.

5.3.7. Pete wants to make a special die that has the shape of a cube with dots on the faces. The number of dots on different faces must be different; moreover, on any two adjacent faces the numbers should differ by at least two. The fact that on some faces there may be more than six dots does not bother Pete. What is the minimal number of dots that can be drawn on such a cube? Give an example and prove that a smaller number of dots is impossible.

Other examples of error correction codes are discussed in the articles [**Fut, K**].

Suggestions, solutions, and answers

5.3.1. (*Olympiad in cryptography*) *Answer*: TITANIC.

Suggestion. Note that crossing the wires does not change the number of ones in the representation of the transmitted letter. Under each letter of the received text we write the column consisting of all letters representation by the code with the same number of ones. For example, under the letter A we put letters B, D, H, P. Next we try to "find" a meaningful word choosing one letter from each column. The only option is the word TITANIC.

5.3.2. (d) *Solution* (Written by D. Ibragimov). Suppose that ten digits suffice. Denote the letter codes by a_1, a_2, ..., a_{26}. Note that there are at most eight different beginnings of length 3. According to the pigeonhole principle, there are three codes a_i, a_j, a_k with the same beginnings of length 3. Then, according to the pigeonhole principle, at least two codes of a_i, a_j, a_k match in the fourth digits. Without loss of generality, we assume that these codes are a_i and a_j.

Denote by a'_s the end interval of length 6 of the code a_s. Then a'_i and a'_j are distinct at least in five digits. The word a'_k is distinct at least in 4 digits from a'_i. Then a'_k and a'_j differ in not more than three digits. Therefore, the codes a_k and a_j differ in not more than four digits. But they differ at least in five digits. This is a contradiction.

5.3.3. The guessing algorithm is based on the use of *Hamming code*. See [**SA**].

5.3.5. (b)*Answer*: 64.

Suggestion [**Bug**]. Obviously, in each column of 8 cubic cells there can be only one rook, so you cannot put more than 64 rooks in the cube.

We show how to place 64 rooks so that they do not threaten each other. We introduce a coordinate system with axes directed along the edges of the cube, so that each cell has coordinates (x, y, z) three numbers between 0 and 7. We put the rooks into cells with the sum of coordinates divisible by 8. This is the required arrangement.

First let us prove that these rooks do not threaten each other. Suppose the opposite. Let two rooks threaten each other. So, two of their coordinates (say x and y) coincide, but the third differs (we denote them by z_1 and z_2, respectively). By construction both $x + y + z_1$ and $x + y + z_2$ are divisible by 8. So their difference $z_1 - z_2$ is also divisible by 8, which is impossible since z_1 and z_2 are different nonnegative numbers less than 8.

Now let us prove that each vertical column contains a rook; that is, we placed 64 rooks. Each such column is defined by its pair of coordinates x and y. The z coordinate for the rooks in this column are uniquely determined by the condition $x + y + z \equiv 0 \pmod 8$. Namely, if $x + y$ is divisible by 8, then $z = 0$; otherwise $z = 8 - ((x + y) \mod 8)$.

5.3.6. (b) See [**SA**].

(c) See [**SA**]. The required set of sequences is called the *Hamming code*.

4. Boolean cube (2)
By A. B. Skopenkov

5.4.1. Arrange several knights on a chessboard so that each threatens four others.

5.4.2. Thirty-three letters of the Russian alphabet are encoded by sequences of zeros and ones. What is the shortest sequence length for which the encoding can be made unique?

5.4.3. (a) For a fixed n, the number $\binom{n}{k}$ is maximal for $k = [n/2]$.

(b) *Best in their own ways.* k schoolchildren participated in a mathematical Olympiad. It turned out that for any two schoolchildren A and B there was a problem that A solved and B did not solve, and there was a problem that B solved but A did not solve. What was the smallest possible number of problems? In other words, find the smallest n for which there is a family of k subsets of an n-element set none of which contains (properly) the other.

5.4.4. There is a board with n light bulbs. Each switch can be connected to some bulbs. When the switch button is pressed, the bulbs connected to it change their state: lighted ones go out, and dark bulbs light up. What is the minimal number of switches that allows us to light any subset of bulbs (bulbs not included in this subset should not be lit)?

5.4.5. On the first day of his reign, the king organizes parties for his n subjects. On the second day, the adviser brings the king a list of the names of some subjects (on the first day this list is unknown). On the third day, the king can choose several parties and send to prison those subjects participating in all of them. What is the smallest number of parties to be organized on the first day so that on the third day the king could send to prison everyone from the list (and no one else)?

Remark. The following important construction is useful (although not required) to solve the above problem (and many others). Let us draw points corresponding to all subsets of some n-element set. Moreover, on the kth *floor* we place the points corresponding to k-elements sets. Connect by segments the pairs of points corresponding to sets obtained from each other by adding one element. Then points connected by a segment lie on adjacent floors. The obtained graph is called the n-*dimensional cube*.

Denote by $\mathbb{Z}_2$ the set $\{0, 1\}$. Introduce addition (modulo 2) by formulas

$$0 + 0 = 1 + 1 = 0 \quad \text{and} \quad 1 + 0 = 0 + 1 = 1.$$

Denote by $\mathbb{Z}_2^n$ the set of strings of length n consisting of 0 and 1, with the operation of elementwise addition modulo 2.

A (nonempty) subset of $L \subset \mathbb{Z}_2^n$ is called a *linear subspace* if $x + y \in L$ for any $x, y \in L$ (not necessarily different). In other words, a *linear subspace* is a family of subsets of n-element sets which together with any two subsets contains their symmetric difference (i.e., the sum modulo 2).

5.4.6. (a) Any linear subspace contains the zero string $(0, \dots, 0)$.
(b) The number of elements in any linear subspace is a power of two.

Denote by $\begin{vmatrix} n \\ k \end{vmatrix}$ the number of linear subspaces in $\mathbb{Z}_2^n$ consisting of 2^k elements (such linear subspaces in $\mathbb{Z}_2^n$ are called k-*dimensional*).

5.4.7. (a) Find $\begin{vmatrix} 2 \\ k \end{vmatrix}$ for $k = 0, 1, 2$.

(b) Find $\left|\begin{array}{c} 3 \\ k \end{array}\right|$ for $k = 0, 1, 2, 3$.

(c) Prove the equalities $\left|\begin{array}{c} n \\ 0 \end{array}\right| = \left|\begin{array}{c} n \\ n \end{array}\right| = 1$ and $\left|\begin{array}{c} n \\ 1 \end{array}\right| = \left|\begin{array}{c} n \\ n-1 \end{array}\right| = 2^n - 1$.

(d) Prove the equality $\left|\begin{array}{c} n \\ k \end{array}\right| = \left|\begin{array}{c} n \\ n-k \end{array}\right|$.

(e) Prove the equality $\left|\begin{array}{c} n+1 \\ k+1 \end{array}\right| = \left|\begin{array}{c} n \\ k+1 \end{array}\right| + 2^{n-k}\left|\begin{array}{c} n \\ k \end{array}\right|$.

(f) Find $\left|\begin{array}{c} n \\ 2 \end{array}\right|$.

(g) Find $\left|\begin{array}{c} n \\ k \end{array}\right|$.

Hints

5.4.1. First, set up several knights so that each threatens *one* other knight. Then set up several knights so that each threatens *two* other knights.

5.4.3. (b) *Answer:* the smallest $n = n(k)$, for which $\binom{n}{[n/2]} \geq k$.

5.4.4. *Answer:* n.

5.4.7. (a) *Answer:* $1, 3, 1$.

(b) *Answer:* $1, 7, 7, 1$.

(d) Use the orthogonal complement.

(f) One can choose an ordered pair of linearly independent vectors in n-dimensional linear space over $\mathbb{Z}_2$ in $(2^n - 1)(2^n - 2)$ ways.

(g) One can choose an ordered set of k linearly independent vectors in n-dimensional linear space over $\mathbb{Z}_2$ in $(2^n - 2^0) \cdot (2^n - 2^1) \cdot \cdots \cdot (2^n - 2^{k-1})$ ways.

Suggestions, solutions, and answers

5.4.1. See hint above. Then arrange several knights so that each threatens *three* others.

5.4.3. (a) *First method.* Induction on n with the use of Pascal's rule proves that for any n, the number $\binom{n}{k}$ as a function of k increases with $k \leq n/2$ and decreases with $k \geq n/2$.[4]

Second method. Consider $\binom{n}{k} \big/ \binom{n}{k+1}$ and use the explicit formula for $\binom{n}{k}$.

(b) For the n-element set X, the family $\binom{X}{\lfloor n/2 \rfloor}$ of all $\lfloor n/2 \rfloor$-element subsets satisfies the property from the formulation of the problem: none of

---

[4]The case $k = n/2$ should be treated slightly differently.

them contains another. It remains to prove that *in any family of subsets of an n-element set none of which contains the other, there are at most $\binom{n}{\lfloor n/2 \rfloor}$ subsets.*

Suggestion for the first solution. For each permutation $(a_1, \ldots, a_n)$ of the set of the first n positive integers, we consider the chain of subsets

$$\{a_1\} \subset \{a_1, a_2\} \subset \cdots \subset \{a_1, \ldots, a_n\} = \{1, \ldots, n\}.$$

In any such chain there is at most one subset of our family. The total number of permutations is $n!$. A subset of a elements is included in the chain for

$$a!(n-a)! \geq \lfloor n/2 \rfloor! \cdot (n - \lfloor n/2 \rfloor)!$$

permutations. Therefore, the number of subsets in this family is at most

$$\frac{n!}{\lfloor n/2 \rfloor!(n - \lfloor n/2 \rfloor)!} = \binom{n}{\lfloor n/2 \rfloor}.$$

Suggestion for the second solution. Consider all subsets of this family S that have the minimal number a of elements. If $a < \lfloor n/2 \rfloor$, then in S we can replace these subsets with the same (or greater) number of $(a+1)$-element subsets, so that in the resulting family none of the sets contains the other.

(Indeed, every a-element subset of S is contained in $n - a$ subsets consisting of $a+1$ elements. None of the last $(a+1)$-element subsets is in S. On the other hand, each $(a+1)$-element subset contains at most $a+1$ subsets consisting of a elements that lie in S. Since $n - a \geq a + 1$, the number of $(a+1)$-element subsets containing a subset from S is at least the number of a-element subsets of S. Then we replace the latter with the former.)

Therefore, we can assume that every set in S has at least $\lfloor n/2 \rfloor$ elements. Similarly, replace the subsets with the maximal number $b > \lfloor n/2 \rfloor$ of elements with $(b-1)$-element subsets. Then we can assume that every set in S has exactly $\lfloor n/2 \rfloor$ elements.

5.4.7. (f) The number of $(k+1)$-dimensional subspaces of the space $\mathbb{Z}_2^{n+1}$

- contained in $\mathbb{Z}_2^n \subset \mathbb{Z}_2^{n+1}$ is equal to $\left|\begin{matrix} n \\ k+1 \end{matrix}\right|$;

- not contained in $\mathbb{Z}_2^n \subset \mathbb{Z}_2^{n+1}$ is equal to $2^{n-k}\left|\begin{matrix} n \\ k \end{matrix}\right|$.

Let us prove the second statement. Note that a $(k+1)$-dimensional subspace L of the space $\mathbb{Z}_2^{n+1}$ not contained in $\mathbb{Z}_2^n$ intersects $\mathbb{Z}_2^n$ in a k-dimensional subspace $L \cap \mathbb{Z}_2^n$. The subspace L is determined by the intersection of $L \cap \mathbb{Z}_2^n$ and by a vector lying in the orthogonal complement to $L \cap \mathbb{Z}_2^n$ in $\mathbb{Z}_2^{n+1}$ but not lying in the orthogonal complement to $L \cap \mathbb{Z}_2^n$ in $\mathbb{Z}_2^n$. There are $2^{n+1-k} - 2^{n-k} = 2^{n-k}$ such vectors.

Remark. The following formula is also valid: $\left|\begin{matrix} n+1 \\ k+1 \end{matrix}\right| = \left|\begin{matrix} n \\ k \end{matrix}\right| + 2^{k+1}\left|\begin{matrix} n \\ k+1 \end{matrix}\right|.$

(h) *Answer:* $\frac{(2^n-1)(2^{n-1}-1)\ldots(2^{n-k+1}-1)}{(2^k-1)(2^{k-1}-1)\ldots(2^1-1)}.$

5. Expressibility for functions of the algebra of logic
By A. B. Skopenkov

Examples and definitions (1)

In this section, letters denote elements of the set $\mathbb{Z}_2 = \{0,1\}$. We introduce the following operations:

logical "not": $\overline{a} = \begin{cases} 1 & \text{for } a = 0, \\ 0 & \text{for } a = 1; \end{cases}$

logical "or": $a \vee b = \begin{cases} 1 & \text{for } a = 1 \text{ or } b = 1, \\ 0 & \text{otherwise}; \end{cases}$

logical "and": $a \& b = \begin{cases} 1 & \text{for } a = 1 \text{ and } b = 1, \\ 0 & \text{otherwise}; \end{cases}$

sum modulo 2, or XOR: $a \oplus b = \begin{cases} 0 & \text{for } a = b, \\ 1 & \text{otherwise}. \end{cases}$

5.5.1. Prove the following equalities:
  (a) $\overline{a \vee b} = \overline{a} \& \overline{b}$;   (b) $(a \vee b) \& c = (a \& c) \vee (b \& c)$;
  (c) $(a \& b) \vee c = (a \vee c) \& (b \vee c)$;   (d) $(a \oplus b) \& c = (a \& c) \oplus (b \& c)$.

5.5.2. Express:
  (a) $x \& y$ through $\overline{x}$ and $\vee$;
  (b) $x \oplus y$ through $\overline{x}$, $\&$, and $\vee$.

Informally speaking, if there are several functions, then some of them can be substituted as arguments of others. This is called *expression* of one function by another. See the formal definition of *superposition* in Section 5 (Post's theorem, below) of this chapter. For example, the function $x^2 y + y^2 + z^2$ is expressed in terms of the functions $x + y$ and xy (i.e., it is a superposition of these functions).

In the process of expressing some functions through others, one variable can be substituted for different variables (i.e., you can consider them as the same). For example, the function $2x$ is expressed in terms of the function $x + y$.

We will often omit $\&$ and write, for example, $(x \& y \& z) \vee (\overline{a} \& b)$ as $xyz \vee \overline{a}b$.

5.5.3. Express, in terms of $\overline{x}$ and $\&$:
  (a) $x \vee y$;
  (b) $x \oplus y$;
  (c) $x | y$, where $1 | 1 := 0$ and $x | y := 1$ if at least one of x, y is 0 (*Sheffer stroke*);
  (d) $x \downarrow y$, where $0 \downarrow 0 := 1$ and $x \downarrow y := 0$ if at least one of x, y is equal to 1 (*Pierce arrow*);
  (e) $x \leq y$, where $(x \leq y) := 1$ if and only if $x \leq y$ "as real numbers".

5.5.4. Functions (mappings) are called *equal* if they take the same value for all values of the arguments. For examples of equal functions, see problem 5.5.1.

(a) How many different functions $f\colon \mathbb{Z}_2^n \to \mathbb{Z}_2$ are there?

(b) Is it true that every function $f\colon \mathbb{Z}_2^2 \to \mathbb{Z}_2$ in two variables can be expressed in terms of $\&$, $\vee$, and $\overline{x}$?

5.5.5. Express through $\&$, $\vee$, and $\overline{x}$ the function $f\colon \mathbb{Z}_2^3 \to \mathbb{Z}_2$ in three variables, which is

(a) equal to 1 on the triple $(0,1,0)$ and equal to 0 on all other triples;

(b) equal to 1 on the triples $(0,1,0)$, $(1,1,1)$ and equal to 0 on all other triples;

(c) an arbitrary function.

5.5.6. Any function $f\colon \mathbb{Z}_2^n \to \mathbb{Z}_2$ can be expressed in terms of

(a) $\&$, $\vee$, and $\overline{x}$;   (b) $\&$ and $\bar{x}$;   (c) $\&$, $\oplus$, and $\mathbf{1}$, where $\mathbf{1}(x) := 1$;

(d) $|$.

5.5.7. Can any function $\mathbb{Z}_q^n \to \mathbb{Z}_q$ be expressed in terms of the sum and product modulo q and constant 1 (i.e., "is" a polynomial, more precisely, corresponds to some polynomial (see [**Sko**, §4.3]) if

(21) $q = 2$, $n = 1$;   (2n) $q = 2$, n is arbitrary;

(31) $q = 3$, $n = 1$;   (3n) $q = 3$, n is arbitrary;

(41) $q = 4$, $n = 1$;   (4n) $q = 4$, n is arbitrary?

5.5.8. (Challenge) Any function $f\colon \mathbb{Z}_2^n \to \mathbb{Z}_2$ is *uniquely* represented by a *Zhegalkin polynomial*, that is, the sum modulo two of *monomials*, i.e., nonrepeating products of variables such that in each monomial all the variables are different. The constant 1 is considered as a monomial with no factors. The constant 0, by definition, is expressed by the empty Zhegalkin polynomial.

5.5.9. Can each function $f\colon \mathbb{Z}_2^n \to \mathbb{Z}_2$ be expressed in terms of

(a) $\oplus$ and 1;   (b) $\vee$ and $\&$;   (c) $\vee$, $\&$, 0, and 1;   (d) $\oplus$ and $\&$;

(e) $f(x,y) = x \oplus y \oplus 1$ and $\vee$;   (f) $g(x,y,z) = xy \vee xz \vee yz$ and $\overline{x}$?

5.5.10. (Challenge) Do there exist operations of "sum" and "product" on $\mathbb{Z}_4$ that satisfy the usual properties (like $x(y+z) = xy + xz$) and such that each function $\mathbb{Z}_4^n \to \mathbb{Z}_4$ "is" a polynomial?

Post's theorem (2*)

If the following definition seems complicated to you, you can skip it. You can try to solve problems 5.5.12–5.5.18 at a less formal level. Therefore, in these, instead of writing "a function is a superposition of functions from F" we write "can be expressed in terms of functions from F".

Suppose we are given some set of functions $F = \{f_\alpha \colon \mathbb{Z}_2^{n_\alpha} \to \mathbb{Z}_2\}_{\alpha \in A}$ (i.e., f_α is a function of n_α variables; the set is not necessarily finite). We define the set $\overline{F}$ of *superpositions* of functions from F as the set of all functions that can be obtained from the functions of the set F and of all the individual variables x_j (i.e., the projections of $\mathbb{Z}_2^n \to \mathbb{Z}_2$ onto the jth coordinate for all different n, j) by a sequence of the following operations (called *elementary superposition*): if functions $f(x_1, \ldots, x_n)$, $g_1(\ldots), g_2(\ldots), \ldots, g_n(\ldots)$ (not necessarily different) are already obtained, then obtain $f(g_1(\ldots), \ldots, g_n(\ldots))$.

Here, as arguments of the functions g_i, you can take any, including coinciding, variables.

Superposition can also be defined graphically, in the language of schemes. Superpositions is an important subject of mathematics and computer science; see [**BMSCS**] and the references therein.

5.5.11. Will an equivalent definition be obtained if we take only $f \in F$, but not an arbitrary function already obtained?

5.5.12. A function $f \colon \mathbb{Z}_2^n \to \mathbb{Z}_2$ is called *linear* if for some $\varepsilon \in \mathbb{Z}_2$ and a subset $\{i_1, \ldots, i_s\} \subset \{1, \ldots, n\}$ we have $f(x_1, \ldots, x_n) = x_{i_1} + \cdots + x_{i_s} + \varepsilon$ for all $(x_1, \ldots, x_n) \in \mathbb{Z}_2^n$ (i.e., if its Zhegalkin polynomial does not contain monomials of degree greater than 1).

(a) If all the functions in a set are linear, then no nonlinear function can be expressed in terms of them.

(b) All linear functions can be expressed in terms of functions from problem 5.5.9(a).

(c) There is no single linear function in terms of which one can express all linear functions.

5.5.13. A function $f \colon \mathbb{Z}_2^n \to \mathbb{Z}_2$ is called *monotone* if for any two n-tuples $(x_1, \ldots, x_n) \geq (y_1, \ldots, y_n)$ the following equality holds: $f(x_1, \ldots, x_n) \geq f(y_1, \ldots, y_n)$. Here the notation $(x_1, \ldots, x_n) \geq (y_1, \ldots, y_n)$ means that these inequalities hold componentwise, i.e., $x_1 \geq y_1$, $x_2 \geq y_2$, $\ldots$, $x_n \geq y_n$.

(a), (b) Solve analogues of problems 5.5.12(a), (b) for monotone functions (problem 5.5.9(c) instead of 5.5.9(a)).

(c) There are no three monotone functions such that every monotone function can be expressed in terms of them.

5.5.14. A function $f \colon \mathbb{Z}_2^n \to \mathbb{Z}_2$ is said *to preserve* 0 if $f(0, 0, \ldots, 0) = 0$.

(a), (b), (c) Solve analogues of problems 5.5.12(a), (b), (c) for functions preserving 0 (problem 5.5.9(d) instead of 5.5.9(a)).

5.5.15. We define functions that *preserve* 1 similarly.

(a), (b), (c) Solve analogues of problems 5.5.12(a), (b), (c) for functions that preserve 1 (problem 5.5.9(e) instead of 5.5.9(a)).

5.5.16. A function $f\colon \mathbb{Z}_2^n \to \mathbb{Z}_2$ is called *self-dual* if $f(\overline{x_1}, \overline{x_2}, \ldots, \overline{x_n}) = \overline{f(x_1, x_2, \ldots, x_n)}$.

(a), (b) Solve analogues of problems 5.5.12(a), (b) for self-dual functions (problem 5.5.9(f) instead of 5.5.9(a)).

(c) There is a single self-dual function f such that each self-dual function can be expressed in terms of f.

5.5.17. (a) The constants 0 and 1 can be expressed in terms of any nonself-dual function and $\overline{x}$.

(b) The constants 0 and 1 can be expressed in terms of any nonmonotone function and $\overline{x}$.

(c) The function & can be expressed in terms of any nonlinear function, $\overline{x}$, 1, and 0,

(d) **The completeness criterion, or Post's theorem.** Given a set of functions, any function can be expresssed in terms of functions from this set if and only if there are functions in this set (not necessarily different) belonging to the complement to each of the five sets listed above (linear, monotone, preserving 1, preserving 0, and self-dual).

These five sets are called *precomplete classes*. Guess why?

5.5.18. (a) In each of the five precomplete classes there are functions that do not belong to other classes; that is, for any two different precomplete classes A and B there is a function f such that $f \subset A$ and $f \notin B$.

(b) A set of functions is called *complete* if any function can be expressed in terms of functions from this set. Prove that when adding one new function to each of the five precomplete classes we obtain a complete set.

The variable x_1 is called *nonessential* for a function $f\colon \mathbb{Z}_2^n \to \mathbb{Z}_2$ if for any $x_2, \ldots, x_n \in \mathbb{Z}_2$ the following equality holds: $f(0, x_2, \ldots, x_n) = f(1, x_2, \ldots, x_n)$. One can similarly define when each of variables x_i, $i = 2, \ldots, n$, is nonessential.

5.5.19. Find a function of three variables such that each variable is essntial.

5.5.20. Functions $\mathbb{Z}_2^n \to \mathbb{Z}_2$ are called *equivalent* if they become equal after renaming variables and adding and removing nonessential variables. Equivalence classes are called *functions of the algebra of logic*.

(a) How many different functions of the algebra of logic have at most two essential variables?

(b)* The same as in part (a), for three essential variables.

A function of the algebra of logic is called *linear* if some (or, equivalently, any) mapping $f\colon \mathbb{Z}_2^n \to \mathbb{Z}_2$ representing it is linear. Other sets of functions of the algebra of logic (monotone, etc.) are defined in a similar way. Post's theorem also holds for functions of the algebra of logic. It is usually formulated in this way, but this makes it a little less accessible for nonspecialists and beginners.

5.5.21.* Come up with an infinite number of different sets of functions of the algebra of logic, closed with respect to superposition.

Hints

5.5.17. (b) A function is monotone if it is monotone in each variable individually.

Suggestions, solutions, and answers

5.5.4. *Answer*: (a) 2^{2^n}.

    5.5.7. (2n), (3n) Any function $\mathbb{Z}_q^n \to \mathbb{Z}_q$ "is" a polynomial for a prime q. To prove this, first obtain functions that vanish at all but one n-tuple.

    5.5.9. *Answer*: no (to all parts of the problem).

    Suggestion. (b), (d), (f) $f(x) = 1$ cannot be expressed; (b), (e), (f) $f(x) = 0$ cannot be expressed; (c) $f(x) = \overline{x}$ cannot be expressed.

    Consider an expression of each above function in terms of the given ones. All variables in it can be considered coinciding.

    5.5.19. *Answer*: xyz.

    5.5.20. *Answer*: (a) 12.

6. Complexity of summation[5]
By Yu. G. Kydryashov and A. B. Skopenkov

The main results of this section are problems 5.6.7(b) and 5.6.14(d).

Introductory problems (2)

5.6.1.* Let $x_1, \dots, x_{99}$ be 99 real numbers. One can add two of the existing numbers and remember the result. What is the minimal number of additions allowing us to find the following sums:

    (a) $x_1 + \cdots + x_{99}$;   (b) $x_1 + \cdots + x_{66}$ and $x_{34} + \cdots + x_{99}$?

Later in this section, only variables taking values in $\mathbb{Z}_2 = \{0, 1\}$ (except for the problem 5.6.3) will be considered. Starting from the set $x_1, \dots, x_n$ of such variables, one can add (modulo 2) two already existing expressions and remember the result.

5.6.2. What is the minimal number of additions we need to find the following sums:

    (a) $x_1 + \cdots + x_{99}$;   (b) $x_1 + \cdots + x_{66}$ and $x_{34} + \cdots + x_{99}$?

---

[5]The authors thank I. Nikokoshev for useful discussions.

5.6.3. Find a set of sums such that the minimal number of additions required to compute it in the case of real variables is greater than the minimal number of additions required to compute it when the variables are in $\mathbb{Z}_2$.

5.6.4. (a) Any two sums $x_{i_1} + \cdots + x_{i_p}$ and $x_{j_1} + \cdots + x_{j_q}$ of the variables $x_1, \ldots, x_n$ can be computed for not more than $n - 1$ additions.

(b) Any three sums $x_{i_1} + \cdots + x_{i_p}$, $x_{j_1} + \cdots + x_{j_q}$, and $x_{k_1} + \cdots + x_{k_r}$ of variables $x_1, \ldots, x_n$ can be computed in not more than $n + 1$ additions.

5.6.5. Any set of sums $x_{i_1} + \cdots + x_{i_p}$ of variables $x_1, \ldots, x_n$ can be computed in not more than $2^n - (n + 1)$ additions.

5.6.6. Suppose any set of m sums of n variables can be found in l additions. Then

(a) any set of km sums of the same n variables can be found in not more than kl additions;

(b) any set of m sums of kn variables $x_1, \ldots, x_{kn}$ can be found in not more than $k(l + m)$ additions.

5.6.7. (The main problem) (a) For sufficiently large m, n, any set of m sums of n variables can be found in at most $\frac{mn}{100}$ additions.

(b) Any set of m sums of n variables can be found in at most $2m \left\lceil \frac{n}{\lfloor \log_2 m \rfloor} \right\rceil$ additions. Here $\lfloor x \rfloor$ is the integer part of x, i.e., the largest integer not exceeding x, and $\lceil x \rceil$ is the upper integer part of the number x, i.e., the smallest integer that is larger than or equal to x.

Definitions and examples (3*)

A sequence of additions can be represented in the following way. We have a set of adders ("mod 2 summators") with two inputs and one output equal to the sum modulo 2 of inputs (this output can be "duplicated"). We need to assemble a circuit of adders that implements the given set of sums.

The number of additions is equal to the number of adders in the circuit. This number is called the *circuit complexity*.

The minimal complexity of the circuit that implements the given set M of sums is called the *complexity of the set of sums* and is denoted by $L(M)$.

In this notation, in problem 5.6.2 we need to find $L(\{x_1 + \cdots + x_{99}\})$ and $L(\{x_1 + \cdots + x_{66}, x_{34} + \cdots + x_{99}\})$.

The minimal number l such that any set of m sums of variables $x_1, \ldots, x_n$ can be found using l adders is called *the complexity of the ordered pair* (m, n) and is denoted by $L(m, n)$.

In this notation, problem 5.6.4 states that $L(2, n) \leq n - 1$ and $L(3, n) \leq n + 1$, and problem 5.6.7(b) states that

$$L(m, n) \leq 2m \left\lceil \frac{n}{\lfloor \log_2 m \rfloor} \right\rceil.$$

A set of m sums of n variables can be represented by a matrix M of size $m \times n$, where at the intersection of the ith row and jth column we put 1 if the jth variable occurs in the ith sum, and 0 otherwise. Then the ith sum can be written very simply as follows:

$$s_i = \sum_{j=1}^{n} M_{ij} x_j.$$

In this notation, the sets of sums from problem 5.6.2 is represented as

$$\begin{pmatrix} 1 & \cdots & 1 \end{pmatrix}, \quad \begin{pmatrix} 1 & \cdots & 1 & 1 & \cdots & 1 & 0 & \cdots & 0 \\ 0 & \cdots & 0 & 1 & \cdots & 1 & 1 & \cdots & 1 \end{pmatrix}.$$

Let us consider a family of m sums of variables $x_0, \ldots, x_{2^m-1}$ where all variables occur "differently". Columns of this matrix are binary representations of integers between 0 and $2^m - 1$. Formally, let $j = b_{m-1,j} 2^{m-1} + \cdots + b_{0,j} 2^0$ be the binary representation of the number $j < 2^m$. Then $B_m = (b_{i,j})$. For example,

$$B_1 = \begin{pmatrix} 0 & 1 \end{pmatrix}, \quad B_2 = \begin{pmatrix} 0 & 0 & 1 & 1 \\ 0 & 1 & 0 & 1 \end{pmatrix}, \quad B_3 = \begin{pmatrix} 0 & 0 & 0 & 0 & 1 & 1 & 1 & 1 \\ 0 & 0 & 1 & 1 & 0 & 0 & 1 & 1 \\ 0 & 1 & 0 & 1 & 0 & 1 & 0 & 1 \end{pmatrix}.$$

For a matrix M with elements M_{ij} we denote by M^t the matrix with elements $(M^t)_{ij} = M_{ji}$. The columns of the matrix M^t are the rows of the matrix M.

5.6.8. The following inequalities hold:
   (a) $L(B_m) < 2^{m+1}$;   (b) $L(m, n) < n + 2^{m+1}$.

5.6.9. Consider a matrix M of size $m \times n$ that has no null rows and no null columns.
   (a) Is it true that $L(M) + m = L(M^t) + n$?
   (b) Is it true that $L(m, n) + m = L(n, m) + n$?
   (c) Find $L(B_m)$.

To conclude this subsection, we present a research problem.

5.6.10.* Consider the set of variables $x_{\alpha_1, \ldots, \alpha_n}$, numbered by the vertices of the n-dimensional unit cube $\mathbb{Z}_2^n$. It is required to find the sums of variables for all faces containing the vertex $(0, \ldots, 0)$. In other words, for each set $(\alpha_1, \ldots, \alpha_n) \in \mathbb{Z}_2^n$ it is needed to find $\sum_{\forall i:\, \beta_i \leq \alpha_i} x_{\beta_1 \ldots \beta_n}$. What is the minimal number of mod 2 additions allowing us to do this?

Asymptotic estimates (4*)

5.6.11. For sufficiently large n, the following inequality holds:

$$L(n,n) < \frac{n^2}{0.9 \log_2 n}.$$

5.6.12. (a) There are at most $(n+k)^{m+2k}$ circuits of k adders with n inputs and m outputs.

(b) If $k = L(m,n)$, then $mn \le (m+2k)\log_2(n+k)$.

(c) For sufficiently large n we have $L(n,n) \ge \frac{n^2}{5\log_2 n}$.

(d) For any $\varepsilon > 0$ and for sufficiently large n (depending on ε) the following inequality holds:

$$L(n,n) \ge \frac{n^2}{(4+\varepsilon)\log_2 n}.$$

To state the next results, we need some asymptotic notation.

A sequence $a_n > 0$ is *asymptotically not bigger* than a sequence $b_n > 0$ if for any $\varepsilon > 0$ there is only a finite set of indices k such that $a_n > (1+\varepsilon)b_n$. We denote this as follows: $a_k \lesssim b_k$. In this notation, problem 5.6.12(d) takes the following form: $L(n,n) \gtrsim \frac{n^2}{4\log_2 n}$.

5.6.13. If $a_n \lesssim b_n$ and $b_n \lesssim c_n$, then $a_n \lesssim c_n$.

A sequence a_n is *little-o* of a sequence b_n ($a_n \ll b_n$) if $a_n/b_n \to 0$ as $n \to \infty$.

5.6.14. If $\log_2 n \ll f(n) \ll 2^n$ for a function $f \colon \mathbb{Z} \to \mathbb{Z}$, then

(a) $\frac{nf(n)}{2\log_2(nf(n))} \lesssim L(f(n), n)$;

(b) $L(f(n), n) \lesssim \frac{nf(n)}{\log_2 f(n)}$;

(c) $L(f(n), n) \lesssim \frac{nf(n)}{\log_2 n}$;

(d) $\frac{nf(n)}{2\log_2(nf(n))} \lesssim L(f(n), n) \lesssim \frac{nf(n)}{\max(\log_2 n, \log_2 f(n))}$.

(This is the main problem.)

Suggestions, solutions, and answers

5.6.1. *Answer*: 98.

Suggestion. We present general arguments that allow us to solve not only this but also many other similar problems. Obviously, the sum $x_1 + \cdots + x_{99}$ can be found in 98 additions. Suppose we have a circuit of adders that implements our sum. Consider a graph in which there are 99 vertices $P_1, \ldots, P_{99}$. For an adder with the first input $x_{i_1} + \cdots + x_{i_p}$ and the second input $x_{j_1} + \cdots + x_{j_q}$ we draw the edge from the vertex P_{i_1} to the vertex P_{j_1}. It is clear that at every moment the vertices corresponding to the variables included into one of the already calculated sums lie in one connected component. Therefore, the finite graph is connected, and that means that in it there are at least 98 edges; i.e., $a \geq 98$.

(b) *Answer*: 98.

Suggestion. The proof is similar to the proof of (a).

5.6.2. The solution repeats the solution of problem 5.6.1.

5.6.3. As an example, take the following set:

$$\{x_1 + x_2, \ x_1 + x_2 + x_3, \ x_1 + x_2 + x_3 + x_4, \ x_2 + x_3 + x_4\}.$$

Indeed, in the case of variables from $\mathbb{Z}_2$ we can find these sums in the indicated order without computing auxiliary sums, but in the case of real variables we have to compute at least one auxiliary sum.

5.6.4. (a) We will prove the statement by induction on n. Denote by $S(A)$ the sum of all variables indexed by elements of A. Let $S(A_1)$ and $S(A_2)$ be the sums sought for. If all the sets $B_1 := A_1 - A_2$, $B_2 := A_1 \cap A_2$, and $B_3 := A_2 - A_1$ are nonempty, then we calculate the sum in each of these sets using at most

$$|B_1| - 1 + |B_2| - 1 + |B_3| - 1 = |A_1 \cup A_2| - 3 \leq n - 3$$

additions. It remains to find

$$S(A_1) = S(B_1) + S(B_2) \quad \text{and} \quad S_2 = S(B_2) + S(B_3).$$

This can be done with two additions.

One can easily (for example, by exhaustive search) prove that if one of the sets is empty, the number of additions does not increase.

(b) The proof is similar to part (a).

5.6.5. It is easy to see that $2^n - (n + 1)$ is the number of all sums of n variables that have at least two terms. We can successively calculate these amounts: first, the sums with two terms, then the sums with three terms, etc., and, finally, the sum with n terms. As a result of this process, we calculate all the sums of the variables $x_1, \ldots, x_n$.

5.6.6. (a) A set of km sums can be viewed as k sets of m sums. By considering each such set separately, we obtain the required estimate.

(b) Let M be the matrix corresponding to our set of m sums of kn variables. Using vertical "cuts" we cut M into k matrices of size $m \times n$. First, using kl adders we implement the additions corresponding to all resulting matrices of size $m \times n$. Then each sought for sum is the sum of at most $k - 1$ already found sums. Therefore, the required set of sums can be implemented using at most $kl + m(k - 1) < k(l + m)$ additions.

5.6.7. (a) Using problem 5.6.5 with $n = 500$ we see that any set of m sums of 500 variables can be computed using at most $l = 2^{500}$ additions. Therefore, by problem 5.6.6(b) any set of m sums of $500k$ variables can be computed using at most $k(l + m)$ additions. Thus, any set of m sums of n variables can be computed using at most

$$\left\lceil \frac{n}{500} \right\rceil (l + m)$$

additions. It is clear that for sufficiently large m and n this number is less than $mn/100$.

(b) Problem 5.6.5 shows that any set of m sums of $\lfloor \log_2 m \rfloor$ variables can be computed using at most $l = 2^{\lfloor \log_2 m \rfloor} \leq m$ additions. Therefore, by problem 5.6.6(b), any set of m sums of $[\log_2 m]k$ variables can be computed using at most $k(l + m) \leq 2km$ additions. Thus, any set of m sums of n variables can be computed using at most

$$2m \left\lceil \frac{n}{\lfloor \log_2 m \rfloor} \right\rceil$$

additions.

5.6.8. (b) Consider an arbitrary matrix A of size $m \times n$. If in this matrix all columns are different, then the complexity of this matrix is not greater than $L(B_m)$. If there are two matching columns, then, first calculating the sum of the variables corresponding to these columns, we reduce the problem to the implementation of a matrix of size $m \times (n - 1)$. Therefore, $L(m, n) \leq L(B_m) + n < n + 2^{m+1}$.

5.6.9. (a) *Answer*: true.

Suggestion. Construct a circuit in the form of a directed graph that has vertices of four types:[6]

- circuit inputs (they have no inputs and one output),
- adders (they have two inputs and one output),
- splitters (they have one inputs and two outputs),
- circuit outputs (they have one input and no outputs).

It is easy to see that the element M_{ij} is equal to the number of paths from the jth input to the ith output taken modulo 2. Consider a new graph in which all edges change direction to the opposite, and consider the corresponding circuit (so that inputs become outputs and vice versa, and splitters become combiners and vice versa). It is easy to see that this circuit implements the matrix M^t.

Let l_1 be the complexity of the first circuit, and let l_2 be the complexity of the second circuit. We will compute the number of edges of our graph in two ways: as the number of beginnings of edges of the first graph and as the number of beginnings of edges of the second graph. In the first case we get $n + l_1 + 2l_2$, and in the second case we get $m + l_2 + 2l_1$. Therefore, $n + l_2 = m + l_1$. Since $l_1 = L(M)$, we get that $L(M^t) \leq m - n + L(M)$. It remains to use the fact that $(M^t)^t = M$.

(b) *Answer*: true.

Suggestion. This follows from (a).

(c) *Answer*: $L(B_m) = 2^{m+1} - 2m - 2$.

Suggestion. We remove the zero column from the matrix B_m and denote the resulting matrix B'_m. Then $(B'_m)^t$ is the matrix corresponding to the set

---

[6]*Editor's note.* In electrical engineering and in computer science, circuits perform various operations on analog or digital signals; for us, circuits are just directed graphs with vertices of specific form.

of all sums of m variables. We already know the complexity of this set of sums. Using part (a), we get the answer.

5.6.11. Problems 5.6.6(a) and 5.6.8 show that for sufficiently large n the following chain of inequalities holds:

$$\mathrm{L}(n,n) < \left\lceil \frac{n}{\lfloor 0.95 \log_2 n \rfloor} \right\rceil \mathrm{L}(\lfloor 0.95 \log_2 n \rfloor, n) < \frac{n}{0.94 \log_2 n} \mathrm{L}(\lfloor 0.95 \log_2 n \rfloor, n)$$

$$< \frac{n}{0.94 \log_2 n}(n + 2 \cdot 2^{\lfloor 0.95 \log_2 n \rfloor}) < \frac{n}{0.94 \log_2 n}(n + 2n^{0.95}) < \frac{n^2}{0.9 \log_2 n}.$$

5.6.12. (a) Call *wire beginnings* inputs of the circuit and outputs of adders (total $n+k$), and call *wire ends* the outputs of the circuit and inputs of adders (total $m + 2k$). We enumerate all the beginnings and ends of the wires. Let us complement each circuit with information about which wire beginnings are connected to which ends. It is clear that this information uniquely determines the circuit. Therefore, the required number of circuits does not exceed the number of different connections of $m + 2k$ wire ends with some of the $n + k$ wire beginnings, which is equal to $(n + k)^{m+2k}$.

(b) Since the number of different circuits with $\mathrm{L}(m,n)$ elements with n inputs and m outputs does not exceed the number of matrices of size $m \times n$ with elements in $\mathbb{Z}_2$, we have $2^{mn} \leq (n + k)^{m+2k}$. Taking the logarithm of this inequality, we obtain the statement of the problem.

(c) If $k = \mathrm{L}(n,n) < \frac{n^2}{5 \log_2 n}$, then from part (b) it follows that

$$n^2 \leq \left(n + \frac{2n^2}{5 \log_2 n}\right) \log_2 \left(n + \frac{n^2}{5 \log_2 n}\right),$$

which is not true for sufficiently large n, since

$$\lim_{n \to \infty} \frac{1}{n^2}\left(n + \frac{2n^2}{5 \log_2 n}\right) \log_2 \left(n + \frac{n^2}{5 \log_2 n}\right)$$

$$= \lim_{n \to \infty} \frac{\log_2 n}{n^2}\left(n + \frac{2n^2}{5 \log_2 n}\right) \cdot \lim_{n \to \infty} \frac{1}{\log_2 n} \log_2 \left(n + \frac{n^2}{5 \log_2 n}\right)$$

$$= \frac{2}{5} \cdot 2 = \frac{4}{5}.$$

(d) The proof is similar to the previous one.

5.6.13. If $a_n < b_n\sqrt{1 + \varepsilon}$ and $b_n < c_n\sqrt{1 + \varepsilon}$, then $a_n < (1 + \varepsilon)c_n$. It remains to use the definition.

5.6.14. (a) We will denote $m = f(n)$. If

$$k = \mathrm{L}(m,n) < \frac{mn}{(2 + \varepsilon) \log_2(mn)},$$

then by problem 5.6.12(b)

$$mn \leq \left(m + \frac{2mn}{(2 + \varepsilon) \log_2(mn)}\right) \log_2 \left(n + \frac{mn}{2 \log_2(mn)}\right).$$

Let us prove that this inequality is not true for sufficiently large n. Indeed,

$$\lim_{n\to\infty} \frac{1}{mn}\left(m + \frac{2mn}{(2+\varepsilon)\log_2(mn)}\right)\log_2\left(n + \frac{mn}{(2+\varepsilon)\log_2(mn)}\right)$$

$$= \lim_{n\to\infty}\frac{\log_2(mn)}{mn}\left(n + \frac{2mn}{(2+\varepsilon)\log_2(mn)}\right)$$

$$\times \lim_{n\to\infty}\frac{1}{\log_2(mn)}\log_2\left(n + \frac{mn}{(2+\varepsilon)\log_2(mn)}\right)$$

$$= \frac{2}{2+\varepsilon}\cdot 2 = \frac{4}{2+\varepsilon}.$$

It remains to use the definition of limit.

(b) Take an arbitrary number k independent of n. We break the matrix A that we want to implement into submatrices of the width $\left\lfloor \log_2\frac{m}{k}\right\rfloor$. Each submatrix can be implemented using $\left\lceil\frac{m}{k}\right\rceil$ additions. Therefore, all submatrices together can be implemented using $\left\lceil\frac{n}{\left\lfloor\log_2\frac{m}{k}\right\rfloor}\right\rceil\cdot\left\lceil\frac{m}{k}\right\rceil$ additions, and the matrix A itself can be implemented using

$$\left\lceil\frac{n}{\left\lfloor\log_2\frac{m}{k}\right\rfloor}\right\rceil\cdot\left\lceil\frac{m}{k}\right\rceil + m\left\lceil\frac{n}{\left\lfloor\log_2\frac{m}{k}\right\rfloor}\right\rceil \lesssim \frac{k+1}{k}\frac{nm}{\log_2 m}$$

additions. Therefore, for any k the following asymptotic inequality holds:

$$\mathrm{L}(m,n) \lesssim \frac{k+1}{k}\frac{mn}{\log_2 m}.$$

Therefore, $\mathrm{L}(m,n) \lesssim \frac{mn}{\log_2 m}$.

(c) Similarly to problem 5.6.11, it can be proved that for any $\varepsilon \in (0,1)$ for sufficiently large n the following inequality holds:

$$\mathrm{L}(m,n) \leq \frac{mn}{\varepsilon\log_2 n}.$$

(d) Follows from the three previous problems.

Chapter 6

Probability
By A. B. Skopenkov
and A. A. Zaslavsky

This chapter[1] is devoted to the simplest concepts and applications of probability theory. To understand it, you need to be familiar with the basics of combinatorics, for example, Sections 1 and 3 in Chapter 1 of this book. Additionally, it is helpful to encounter "physical" interpretations of probability theory; see, for example, [**KZhP**]. Here we immediately start with "mathematical" definitions. However, we present many problems informally and show by examples how to formalize them. We leave the formalization of some problems to the reader. See, for example, problems 6.2.6(c), (d) and 6.4.13, [**GDI2**, problems 6.3.1.b and 6.3.3.c].

1. Classical definition of probability (1)

Consider an experiment with m equally possible outcomes, such as throwing a die, pulling a card from a deck, etc. If the event we are interested in (for example, getting six, drawing an ace, etc.) occurs in a of these outcomes, then the *probability* of the event is considered equal to $p = a/m$.

This explanation is useful for a beginner, but it is not a mathematical definition. Here is a mathematical definition.

The *probability* of a subset A of a finite set M is defined to be the ratio

$$P(A) = P_M(A) := |A|/|M|.$$

Unless otherwise stated, the set M is fixed and is omitted from the notation. Then the probability is defined for all its subsets. These are often called *events*.

6.1.1. From a deck of 52 cards, one card is pulled. Find the probability that it will be
    (a) black;   (b) an ace;   (c) a face card;
    (d) the queen of spades;   (e) a king or a diamond.

---

[1]The author is grateful to Yu. N. Tyurin and T. Takebe for a useful discussion.

For example, in problem 6.1.1(c) the set M ("of all possible outcomes") coincides with the set of cards in the deck, and the set A ("outcomes in which the event in question occurs") coincides with the subset of face cards. So this and many other probabilistic problems can be formulated rigorously in combinatorial terms.

6.1.2. A coin is flipped 3 times. Find the chance of
    (a) three heads;   (b) two heads and one tail.

6.1.3. Find the probability that when rolling two dice,
    (a) the first die shows a larger number of dots than the second;
    (b) the total number of dots will be $2, 3, \ldots, 12$.

6.1.4. Find the probability that a random integer from 1 to 105, inclusive
    (a) is divisible by 5;   (b) is divisible by 7;   (c) is divisible by 35.
    (a'), (b'), (c') Same question, but for a random integer from 1 to 100, inclusive.

6.1.5. Fred knows the answers to 10 questions out of 30. A quiz consists of two questions. What is the probability that Fred answers both questions?

To solve some of the above problems, the following rules are useful.

6.1.6. (a) **Rule of addition.** Let $A \cap B = \emptyset$. Express $P(A \cup B)$ in terms of $P(A)$ and $P(B)$.
    (b) Express the probability $P(A \cup B)$ in terms of $P(A)$, $P(B)$, and $P(A \cap B)$.
    (c) **Multiplication rule.** Express the probability of $P_{M \times N}(A \times B)$ in terms of $P_M(A)$ and $P_N(B)$.
    Comment: $P_M(A) = P_{M \times N}(A \times N)$ and $P_N(B) = P_{M \times N}(M \times B)$.

6.1.7. (a) Red and black socks lie in the drawer. What is the minimum number of socks in the drawer if the probability that two randomly selected socks are red equals $1/2$?
    (b) The same question as above, but it is known that the number of black socks is even.

6.1.8.* (a) A triangle is formed by selecting three vertices of a regular $2n$-gon randomly. What is the probability that it will be right? Acute? Obtuse?
    (If you cannot solve the problem, then see the next section.)
    (b) Find the limits of the probabilities obtained for $n \to \infty$. (Think about the meaning of the results. Compare with problem 6.2.6(c).)

Suggestions, solutions, and answers

6.1.4. *Answers*: (a) 0.2; (b) $\frac{1}{7}$; (c) $\frac{1}{35}$; (a') 0.2; (b') 0.14; (c') 0.05.

(a) *Solution* (E. Pavlov). Let $M = \{1, 2, \ldots, 105\}$ be the set of all possible outcomes, and let $A = \{5, 10, \ldots, 105\} = \{x \in M : 5 \mid x\}$ be the set of favorable outcomes. Then, by definition, the probability of the set A is $P(A) = \frac{|A|}{|M|} = \frac{\lfloor \frac{105}{5} \rfloor}{105} = 0.2$.

6.1.5. *Answer*: $\frac{3}{29}$.

Solution (P. Belopashentseva). Let M be the set of all (nonordered) pairs of distinct numbers from 1 to 30. This set corresponds to the set of all possible quizzes. The number of elements in M is $|M| = \binom{30}{2}$.

Denote by A the set of all nonordered pairs of different numbers from 1 to 10. The set A corresponds to the set of quizzes for which Fred knows both answers. The number of elements in A is $|A| = \binom{10}{2}$. The probability of a subset A in a set M is by definition $P_M(A) = |A|/|M| = \binom{10}{2}/\binom{30}{2} = \frac{10 \cdot 9/2}{30 \cdot 29/2} = \frac{3}{29}$.

6.1.6. *Answers*: (a) $P(A \cup B) = P(A) + P(B)$;

(b) $P(A \cup B) = P(A) + P(B) - P(A \cap B)$;

(c) $P_{M \times N}(A \times B) = P_M(A) P_N(B)$.

6.1.7. *Answers*: (a) 4; (b) 21.

Suggestion. Let r and b be the number of red and black socks, respectively. Then the probability of drawing two red socks is $\frac{r(r-1)}{(r+b)(r+b-1)}$. This expression is $1/2$ for an infinite set of pairs (r, b), the smallest of which are $(3, 1)$ and $(15, 6)$. For more details, see Problem 1 from [**Mos**].

6.1.8. (a) *Answer*: $\frac{3}{2n-1}$, $\frac{n-2}{2(2n-1)}$, $\frac{3(n-2)}{2(2n-1)}$.

Suggestion. Consider the first vertex to be fixed. If the second vertex is directly opposite to the first, then the triangle will definitely be right. Otherwise, the third vertex must be at one of the vertices opposite (with respect to the center of the polygon) to one of the first two. Therefore the probability of a right triangle is $\frac{1}{2n-1} + \frac{2n-2}{2n-1} \frac{2}{2n-2} = \frac{3}{2n-1}$.

Similarly, a triangle will be acute if the second vertex is not opposite to the first, and the third vertex lies "between" the vertices that are opposite to the first two. The probability of this is

$$\frac{2}{2n-1} \left(\frac{1}{2n-2} + \frac{2}{2n-2} + \cdots + \frac{n-2}{2n-2} \right) = \frac{n-2}{2(2n-1)}.$$

(b) *Answer*: 0, 1/4, 3/4.

2. A more general definition of probability (1)

6.2.1. (a) One shooter hits a target with probability 0.8, the other with probability 0.7. Find the probability of hitting the target if both shoot simultaneously.

(In this and some other problems below, we delay providing precise formalizations.)

(b) A worker uses three machines. The probabilities that each machine stops is, respectively, equal to 0.1, 0.2, and 0.15. Find the probability that all the machines are working.

To formalize the above problems, it is necessary to add a more general definition. Let a set M be given, and let each $m \in M$ be associated with a nonnegative number $P(m)$, where the sum of $P(m)$ over all $m \in M$ is 1. Then we define the *probability* of the event $A \subseteq M$ to be the sum of the numbers $P(m)$ over all $m \in A$.

For example, in the above problem, it is reasonable to assume that the set M consists of four elements: both shooters hit, the first hits and the second misses, the first misses and the second hits, both miss.

6.2.2. Formulate and prove the analogues of the rules of sum and product for the above generalization.

6.2.3. A father, mother, and son like to play chess. The father promises his son a prize if he wins two games in a row out of three played alternately with the father and the mother. The son knows that the father plays better than the mother. Would it be more advantageous for the son to play the first game with the father or the mother?

The above definition can be generalized to the case of an infinite set M. (In this case, for all $m \in M$, except for a countable number, $P(m) = 0$.) Even more interesting is the following generalization.

6.2.4. Find the probability that a random point of an equilateral triangle lies
   (a) in the triangle formed by midlines;
   (b) in the inscribed circle.

Let $A \subset M$ be subsets of a line (or plane or space) that have a length (or area or volume). Not all subsets have a length (or area or volume); see the remark in Section 5 of Chapter 7. Then the ratio

$$P(A) = P_M(A) := L(A)/L(M),$$

where $L(A), L(M)$ are lengths of the subsets, is called the *probability* of the subset A in M.

Let $A \subset M$ be subsets of a plane (or space) having an area (or volume). Then the ratio

$$P(A) = P_M(A) := S(A)/S(M),$$

where $S(A), S(M)$ are the areas of subsets, is called the *probability* of the subset A in M.

As in the discrete case, when the set M is fixed, the subsets having a length (area, volume) are often called *events*.

6.2.5.* Formulate and prove the analogues of the rules of sum and product for the above "geometric" probabilities.

6.2.6.* (a) Duels in the city of Caution rarely end in a sad outcome. The custom is that each duelist arrives at the place of battle at a random time between 5 and 6 o'clock in the morning and, after waiting for the opponent for 5 minutes, leaves. If the opponent shows up before the 5 minutes end, then an actual duel takes place. What fraction of the duels really ends in a duel?

(b) A stick is randomly broken into three parts. What is the probability that a triangle can be formed from these parts?

(c) Find the probability that a random triangle is acute.

(d) With what probability is a randomly chosen chord in a circle longer than the side of an equilateral triangle inscribed in this circle?

Surprisingly, 6.2.6(c) and (d) have other natural formalizations that give a different answer! It is interesting that parts (c) and (d) allow different natural formalizations which result in different answers!

Suggestions, solutions, and answers

6.2.1. (a) *Answer*: $1 - (1 - 0.7)(1 - 0.8) = 0.94$.

6.2.3. *Answer*: with father.

Solution. Let p_1 and p_2 denote the probability of winning one game from the father and from the mother, respectively: $0 < p_1 < p_2$. Let M be the set of strings of length 3 using the symbols 0 and 1 (for each $k = 1, 2, 3$ the symbol at the kth place "encodes" the result of the kth game). Let $P_1(m)$ and $P_2(m)$ be the probabilities of the element $m \in M$ in the case when the first game is played with the father or the mother, respectively. These probabilities are determined by the product rule (see (6.2.2)). In particular, $P_1(111) = p_1 p_2 p_1$, $P_1(110) = P_1(011) = (1 - p_1)p_2 p_1$, $P_2(111) = p_2 p_1 p_2$, $P_2(110) = P_2(011) = (1 - p_2)p_1 p_2$. Let $A = \{111, 110, 011\}$. Then by definition,

$$P_1(A) = P_1(111) + P_1(110) + P_1(011) = p_1 p_2 (2 - p_1) > p_2 p_1 (2 - p_2) = P_2(A).$$

6.2.6. (a) *Answer*: $23/144$.

(b) For M one can take an equilateral triangle with a height equal to the length of the stick. Since for each point inside the triangle, the sum of the distances from it to the sides is equal to the height, these distances can be considered equal to the lengths of the parts of the rod obtained during the fracture.

For each of these parts to be less than the sum of the other two means that the point representing the break in the equilateral triangle should lie in the equilateral triangle formed by the midlines. Therefore the desired probability is $1/4$.

(c) *First interpretation. Answer*: $12 \ln 2 - 8$. A triangle composed of the resulting pieces can be associated with each break of the stick from part (b). Moreover, the resulting triangle is acute-angled if and only if the sum of the squares of the lengths of any two pieces is greater than the square of the length of the third.

Second interpretation. Answer: $1/4$. With each break of the stick from (b), we can associate an auxiliary triangle, whose *angles* are proportional to the lengths of the resulting pieces. Moreover, a triangle can be made up of such pieces if and only if the auxiliary triangle is acute-angled. Therefore, it is natural to assume that the desired probability is equal to the value obtained in part (b).

(d) If we assume that the endpoints of the chord are uniformly distributed on a circle, we obtain $1/3$. If the middle of the chord is uniformly distributed in the disk, then we obtain $1/4$. And if we draw the diameter perpendicular to the chord and select the intersection point with the chord uniformly on the diameter, then the answer is $1/2$.

3. Independence and conditional probability (1)

The following definition generalizes the notion of the multiplication rule 6.1.6(c). Subsets (i.e., events) A and $B \neq \emptyset$ of a finite set M are *independent* if the fraction (i.e., probability) of the set $A \cap B$ in B is equal to the fraction of the set A in M. We shall give a symmetric reformulation that works also for $B = \emptyset$. The subsets A and B of a finite set M are called *independent* if

$$|A \cap B| \cdot |M| = |A| \cdot |B|.$$

The main example of independent subsets: In the set of all squares of a chessboard, consider the subset formed by the first three rows and the subset formed by the last four columns. More rigorously, consider $A \times N$ and $M \times B$ in $M \times N$.

6.3.1. Subsets that are not independent are called *dependent*. Are the following subsets dependent?

(a) Subsets $\{1, 2\} \subset \{1, 2, 3, 4\}$ and $\{1, 3\} \subset \{1, 2, 3, 4\}$.

(b) Subsets $\{1, 2\} \subset \{1, 2, 3, 4, 5, 6\}$ and $\{1, 3\} \subset \{1, 2, 3, 4, 5, 6\}$.

6.3.2. Are the following subsets of the set of integers from 1 to 105 dependent?

(a) The multiples of 5 and the multiples of 7.

(b) The multiples of 15 and the multiples of 21.

(c) The multiples of 15 and the multiples of 5.

(d) The multiples of 10 and the multiples of 7.

The following reformulation also works for a more general definition of probability when not all numbers $P(m)$ are equal.

The subsets A and B of the set M are called *independent* if $P(A \cap B) = P(A) \cdot P(B)$.

6.3.3. Two noblemen from the King's retinue, awaiting the appearence of His Majesty, decided to play dice. They made the same bets and agreed that the one who first won 10 games gets all the money. With the score 9 to 8, the king appeared and the game had to be discontinued. How should they split the money?

This was one of the problems that laid the foundation of probability theory. (It will be easier for you to solve after looking at problem 6.3.10.)

The problem was posed to the great 17th-century French mathematician Blaise Pascal by an acquaintance, one of those noblemen mentioned in the problem. Pascal realized that the money should be divided in proportion to the chances that the players had for the final victory at the time the game stopped. He found a method for calculating these chances (for any score). Another method for solving the problem, leading to the same result, was found by Pierre Fermat, another great mathematician of the 17th century. Their methods are based on the following concept.

The ratio

$$P(A|B) := P(A \cap B)/P(B)$$

is called the *conditional probability* of the subset A under the condition of the subset B, assuming that $P(B) \neq 0$.

It is clear that the independence of the subsets A and B is equivalent to the fact that $P(A|B) = P(A)$.

6.3.4. (a) A die is tossed, and it is known to be even. Find the probability that it is less than 5.

(b) **Paradox of a boy and a girl.** There are two children in a family. It is known that one of them is a boy. Find the likelihood that the second child is also a boy. (We suppose that the probabilities of giving birth to a boy and girl are equal and that the gender of the second child does not depend on the gender of the first.)

6.3.5. Light bulbs are produced by two factories, the first of which produces 70% of all bulbs. Bulbs produced by the first factory work with probability 0.98, and those produced by the second one work with probability 0.95. Find the probability that the purchased light bulb works.

The solution to this problem is generalized as follows.

6.3.6. The formula for the total probability. If $M = B_1 \cup \cdots \cup B_n$, $B_i \cap B_j$ for $i \neq j$, and $P(B_j) \neq 0$ for all j (they say that $B_1, \ldots, B_n$ is *complete system of events*), then

$$P(A) = P(A|B_1)P(B_1) + \cdots + P(A|B_n)P(B_n).$$

6.3.7. The winner in a match between two boxers is determined by a majority vote of three judges. Two judges make the right decision with probability p, and the third one votes by throwing a coin. Find the probability that these judges make the correct decision.

6.3.8.* The rules of a game of *craps* are as follows: a player rolls two dice. She wins if the sum is 7 or 11 and loses if it is 2, 3, or 12. In the remaining cases, she rolls the dice until she wins by getting the sum she had at the first roll, or she loses with a sum of 7. Find the probability of winning.

6.3.9. Light bulbs are produced by two factories, the first of which produces 70% of all bulbs. Bulbs produced by the first factory work with probability 0.98, and those produced by the second work with probability 0.95. The purchased bulb turned out to be defective. Find the probability that it was produced by the first factory.

The solution to this problem is generalized as follows.

6.3.10. Bayes's formula. The following equality is true:

$$P(B|A) = P(A|B)P(B)/P(A).$$

We often use this corollary of the formulas of 6.3.6 and 6.3.10:

$$P(X|A) = \frac{P(A|X)P(X)}{P(A|B_1)P(B_1) + \cdots + P(A|B_n)P(B_n)}.$$

6.3.11. The probability that a product is defective is 0.04. If it is defective, then it will pass a quality test with probability 0.05, and otherwise with probability of 0.98. Find (to the nearest 0.0001) the probability that a product that passes the test twice is defective.

6.3.12. King Arthur holds a knight elimination tournament. Among the 2^n equally skilled knights, there is a pair of twins. Find the probability that they meet.

6.3.13.* The discriminating bride. (Challenge) A girl chooses a husband for herself from among n applicants, who successively propose to her. Each time she can accept the offer (then everything ends) or reject it (the rejected suitor doesn't propose again). She wants to maximize the probability of choosing the most worthy suitor.

(a) Prove that the optimal strategy is to reject the first $s(n)$ offers and then accept the first one from an applicant superior to all previous ones.

(b) Determine the optimal value of $s(n)$.

Suggestions, solutions, and answers

6.3.3. *Answer*: the first should take 3/4 money, and the second takes 1/4.

Suggestion (N. Medved). Let's see how the game could continue (we believe that the game is fair). With probability 0.5, the first one wins immediately. Otherwise with probability 0.5, the score becomes 9 : 9. In this case, each player has additional probability 0.25 of winning. This means that the odds of winning are in the ratio 3 : 1. Therefore, the first must take three quarters of the money, and the second one takes a quarter.

6.3.4. (b) *First interpretation. Answer*: 1/3.

Suggestion (N. Medved). By the definition of conditional probability, the probability that the second child is a boy, provided that one of the children is a boy, is equal to the probabilities that both children are boys, divided by the probability that at least one of them is a boy. The probability of having two boys is 1/4. The probability of a boy and a girl is 1/2. Hence, the probability that at least one of the two children is a boy is $1 - 0.25 = 0.75$. We have $\frac{0.25}{0.75} = \frac{1}{3}$.

Second interpretation. Answer: 1/2.

6.3.5. *Answer*: $0.7 \cdot 0.98 + 0.3 \cdot 0.95 = 0.971$.

6.3.9. *Answer*: 14/29.

6.3.11. *Answer*: 0.0001.

Suggestion (N. Medved). For a nondefective product, the probability of passing the test twice is 0.9604, and for a defective product, it is 0.0025. Then for an arbitrary product, the probability that it passes the test twice, according to the law of total probability, is equal to

$$0.96 \cdot 0.9604 + 0.04 \cdot 0.0025 = 0.921984 + 0.000100 = 0.922084.$$

According to Bayes's formula (see (6.3.10)), the probability that a product is nondefective, provided that it has passed 2 tests, is equal to the probability that a nondefective product will pass 2 tests, multiplied by the probability of a nondefective product and divided by the probability of passing two tests, that is, $0.9604 \cdot 0.96/0.922084$, or about 99.99% (the precises estimation of error we leave to the reader).

Here is a general argument.

Let A_1 be the event that the product is nondefective, and let A_2 be the event that it is defective. Let B_1 be the event that a positive result was given by the first test, and let B_2 be the event that a positive result was given by the second test. By the conditions of the problem, $P(B_1|A_1) = P(B_2|A_1) = 0.98$, and $P(B_1|A_2) = P(B_2|A_2) = 0.05$. Since the tests are independent, we have $P(B_1 \cap B_2|A_1) = P(B_1|A_1) \cdot P(B_2|A_1)$ and $P(B_1 \cap B_2|A_2) = P(B_1|A_2) \cdot P(B_2|A_2)$. Then by Bayes's formula (see (6.3.10)) this implies $P(A_1|B_1 \cap B_2) = P(B_1 \cap B_2|A_1)P(A_1)/P(B_1 \cap B_2)$.

Since A_1 and A_2 form a complete system of events, expressing $P(B_1 \cap B_2)$ with the help of the full probability formula (see 6.3.6), we find

$$P(A_1|B_1 \cap B_2) = \frac{P(B_1 \cap B_2|A_1)P(A_1)}{P(B_1 \cap B_2|A_1)P(A_1) + P(B_1 \cap B_2|A_2)P(A_2)}$$

$$= \frac{0.98 \cdot 0.98 \cdot 0.96}{0.98 \cdot 0.98 \cdot 0.96 + 0.05 \cdot 0.05 \cdot 0.04} \approx 0.9999.$$

6.3.12. *Answer*: $\frac{1}{2^{n-1}}$.

Suggestion. With probability $\frac{2^{n-1}}{2^n-1}$, the twins are in different halves of the tournament bracket and can only meet in the finals. From here, by induction, we get that the probability of a meeting is $\frac{1}{2^{n-1}}$. See Problem 16 from the book [**Mos**] for more details.

6.3.13. (a) Let the kth suitor be better than all the previous ones. Then the probability that he is the best of all suitors is equal to $\frac{k}{n}$, which is an increasing function of k. On the other hand, it is obvious that the probability of choosing the best suitor after rejecting the kth is a decreasing function of k. Therefore, while the first probability is less than the second, the suitors must be rejected, and when the first probability becomes greater, it is necessary to accept the offer of the first suitor who surpasses all the previous ones.

(b) If the bride acts according to the above strategy, then she receives the best suitor under the following two conditions: the index k of the suitor is more than $s = s(n)$ and the best of the first $k-1$ suitors lies among the first s. The probability of this is $\frac{1}{n}\left(1 + \frac{s}{s+1} + \cdots + \frac{s}{n-1}\right)$, which is approximately equal to $\frac{s}{n}\ln(n/s)$. Therefore, the optimal value s is approximately equal to $\frac{n}{e}$. Moreover, the probability of choosing the best groom for large n is approximately equal to $\frac{1}{e}$ (see [**Mos**, Problem 47]).

4. Random variables (3)

Let M be a finite or countable set and for each element $m \in M$, let $P(m) \geq 0$ be a number (probability). Let us assume that $\sum_{m \in M} P(m) = 1$. A function X defined on M is called a *random variable*. Define the *mass function* of the random variable X to be the set of pairs $(x_i, p_i), i = 1, 2, \ldots,$ where $\{x_1, x_2, \ldots\}$ is the set of possible values of the random variable X and $p_i = P(\{m \in M: X(m) = x_i\})$, $i = 1, 2, \ldots,$ are the corresponding probabilities.

Comment. As a rule, when studying a random variable X it is not necessary to know on which set it is defined. It is sufficient to just know its mass function.

We shall abbreviate the event $\{m \in M: X(m) = x_i\}$ by $X = x_i$.

6.4.1. A coin is flipped 5 times. Find the mass function of the number of heads.

6.4.2. (a) You are invited to play the following game. You pay 2 candies and then roll a die, and you get as many candies as the number rolled. Is this game profitable for you?

(b) Same game, only now if you roll a 1, you must pay 100 candies. (You have enough candies to pay.) Is this game profitable for you?

(c) A bank offers you the following deal. You put 8 candies in the bank, after which a die is rolled. If you roll a 2, 3, or 4, then you get back your contribution plus one additional candy. If you roll 5 or 6, then you receive your contribution plus two additional candies. But if you roll a 1, then you lose your contribution. Is this game profitable for you?

The *expectation* or *average* of the random variable X is the sum

$$E(X) = \sum_i x_i p_i = x_1 P(X = x_1) + x_2 P(X = x_2) + \cdots .$$

Comment. If the set of values of a random variable is infinite, this definition needs to be clarified. The sum of the series in the right part is called the *expectation* only if this series converges absolutely. Otherwise, we say that the variable X *does not have an expectation*. For example, let a random variable X take the value $n \in \mathbb{N}$ with probability $p_n = \frac{1}{n(n+1)}$. Then the series $\sum n p_n = \sum \frac{1}{n+1}$ diverges; i.e., $E(X)$ does not exist. In this chapter, we will assume that all the random variables that we consider have expectations; i.e., the series $\sum x_i P(X = x_i)$ converges absolutely.

6.4.3. (a) Prove that the expectation of the random variable X, defined on the set M, is equal to $\sum_{m \in M} X(m) P(m)$.

(b) Prove that if $E(X) \le x$, then there exists $m \in M$ such that $X(m) \le x$.

(c) Let the random variable X take the same value μ: $X(m) = \mu$ for all $m \in M$. Find $E(X)$.

(d) Express $E(aX + bY)$, where a, b are real numbers and X, Y are random variables, in terms of a, b, $E(X)$, $E(Y)$.

(e) Is it possible to express $E(XY)$ in terms of $E(X)$ and $E(Y)$?

Random variables X and Y are called *independent* if the events $X = x_i$ and $Y = y_j$ are independent for any x_i, y_j; i.e.,

$$P(\{m \in M : X(m) = x_i \text{ and } Y(m) = y_j\}) = P(X = x_i)P(Y = y_i).$$

Informally, independence means that the values of one of the random variable do not affect the mass function of another.

6.4.4. Prove that if the random variables X and Y are independent, then the expectation of their product is equal to the product of their expectation: $E(XY) = E(X)E(Y)$.

The *variance* of a random variable X is the quantity $\mathrm{Var}(X) = E\big((X - E(X))^2\big)$.

Comment. If the set of values of a random variable is infinite, then the variance may not exist. In this chapter, we assume that for any random variable under consideration, the variance exists.

6.4.5. Prove that $\mathrm{Var}(X) = E(X^2) - E(X)^2$.

6.4.6. Prove that if X and Y are independent, then $\mathrm{Var}(X+Y) = \mathrm{Var}(X) + \mathrm{Var}(Y)$.

6.4.7. Chebyshev inequality. Prove that for any random variable X and any $\varepsilon > 0$ the following inequality holds:

$$P(|X - E(X)| \geq \varepsilon) \leq \mathrm{Var}(X)/\varepsilon^2.$$

6.4.8. Fred knows the answers to 20 of 30 questions. A quiz contains 3 questions. Find the mass function of the number of questions that Fred can answer.

6.4.9. Two identical decks of cards are shuffled, and cards are sequentially laid out in pairs on the table. Find the average number of pairs for which the two cards are the same.

6.4.10. In the city of N, bosses are required to provide all employees with a day off if at least one of the employees has a birthday on that day. All other days are workdays. How many people should be hired to have the maximal average productivity during the course of a year?

6.4.11. In problem 6.4.8, find the average value of Fred's grade (if Fred answers 3 questions, he will get 5, for 2 he will get 4, etc.).

6.4.12. Consider the following popular gambling game: a player can bet on one number from 1 to 6. Three dice are thrown, and if the selected number appears on at least one die, then the player gets her bet back plus the same amount for each occurrence of the selected number. Is the game profitable for the player?

6.4.13. (Challenge) A field has the shape of a square with side $350\,\mathrm{m}$. When measuring the side, the probability of an error $\pm10\,\mathrm{m}$ is 0.16; of $\pm20\,\mathrm{m}$ it is 0.08; of $\pm30\,\mathrm{m}$ it is 0.05. Find the average value of the measured area.

Comment. Actually, the answer to this question depends on how the concept of measuring area is formalized. If we independently measure each side of the square and multiply the resulting values, then by problem 6.4.4, the expected value will be $350^2\,m^2$. If only one side is measured and this is then squared, the answer is different.

6.4.14. For a fixed k, find the expected number of blocks of k identical digits in a sequence of n ones and m zeros written in random order.

6.4.15. From a deck of 52 cards, cards are pulled out until the first ace is drawn. What is the average number of cards pulled out?

6.4.16. Along a narrow road n cars go in one direction. At the beginning the speeds of all cars are different. Each car travels at a constant speed until it catches up with the car in front of it, after which it rides at the speed of that car. As a result, eventually the cars are divided into several groups. Find the expected number of groups.

Suggestions, solutions, and answers

6.4.3. *Solution* (T. Cherganov). (a)

$$E(X) = \sum_i x_i P(X = x_i) = \sum_i \left(x_i \sum_{m \in (X=x_i)} P(m) \right)$$

$$= \sum_i \sum_{m \in (X=x_i)} X(m)P(m) = \sum_{m \in M} X(m)P(m).$$

(b) Let $X(m) > x$ for any $m \in M$. Then

$$E(X) = \sum_{m \in M} X(m)P(m) > \sum_{m \in M} xP(m) = x.$$

(c) *Answer:* μ.

$$E(X) = \sum_{m \in M} X(m)P(m) = \mu \sum_{m \in M} P(m) = \mu.$$

(d)

$$E(X + Y) = \sum_{m \in M} (X + Y)(m)P(m)$$

$$= \sum_{m \in M} X(m)P(m) + \sum_{m \in M} Y(m)P(m) = E(X) + E(Y).$$

$$E(aX) = \sum_{m \in M} (aX)(m)P(m) = a \sum_{m \in M} X(m)P(m) = aE(X).$$

6.4.4. *Solution* (T. Cherganov).

$$E(XY) = \sum_k z_k P(XY = z_k)$$

$$= \sum_{i,j} x_i y_j P(X = x_i, Y = y_j) = \sum_{i,j} x_i y_j P(X = x_i) P(Y = y_j)$$

$$= \sum_i x_i P(X = x_i) \sum_j y_j P(Y = y_j) = E(X)E(Y),$$

where $z_1, z_2, \ldots$ are all values of the *product* of random variables X and Y. Think about how to justify the second equality (some z_k can be represented as $z_k = x_i y_j$ in several ways!).

6.4.5. *Solution* (T. Cherganov).

$$\mathrm{Var}(X) = E(X - (E(X))^2) = E(X^2 - 2XE(X) + (E(X))^2)$$

$$= E(X^2) - 2E(X)E(X) + (E(X))^2 = E(X^2) - (E(X))^2.$$

6.4.6. *Solution* (T. Cherganov).

$$\mathrm{Var}(X + Y) = E((X + Y)^2) - (E(X + Y))^2$$

$$= E(X^2) + 2E(X)E(Y) + E(Y^2) - (E(X))^2 - 2E(X)E(Y) - (E(Y))^2$$

$$= \mathrm{Var}(X) + \mathrm{Var}(Y).$$

6.4.7. *Solution* (T. Cherganov). Denote $A := \{m \in M \mid |X(m) - E(X)| \geq \varepsilon\}$. Then

$$\mathrm{Var}(X) = E(X - E(X))^2 = \sum_{m \in M} (X(m) - E(X))^2 P(m)$$

$$\geq \sum_{m \in A} (X(m) - E(X))^2 P(m) \geq \varepsilon^2 \sum_{m \in A} P(m)$$

$$= \varepsilon^2 P(|X(m) - E(X)| \geq \varepsilon).$$

6.4.8. *Solution* (T. Cherganov). *Answer*: $\left(0, \frac{6}{203}\right)$, $\left(1, \frac{45}{203}\right)$, $\left(2, \frac{95}{203}\right)$, $\left(3, \frac{57}{203}\right)$.

The set M of quizzes is a set of unordered triples of different numbers between 1 to 30. Therefore, $|M| = \binom{30}{3}$. Let A_0 be the set of unordered triples of different numbers between 21 to 30. Then $P(A_0) = \binom{10}{3}/\binom{30}{3} = \frac{6}{203}$. Let A_1 be the set of unordered triples of distinct numbers for which two belong to $\{21, \ldots, 30\}$ and one belongs to $\{1, \ldots, 20\}$. Then $P(A_1) = 20\binom{10}{2}/\binom{30}{3} = \frac{45}{203}$. Similarly, $P(A_2) = 10\binom{20}{2}/\binom{30}{3} = \frac{95}{203}$ and $P(A_3) = \binom{20}{3}/\binom{30}{3} = \frac{57}{203}$.

6.4.9. *Answer*: 1.

6.4.10. *Answer*: 364 or 365.

Suggestion. The probability that on the given day none of the n workers have a birthday will be equal to $(364/365)^n$. Therefore, the average number of person-days is $365n(364/365)^n$. This expression reaches its maximal value when n is equal to 364 or 365 (see Problem 34 in [**Mos**]).

6.4.11. *Solution* (T. Cherganov). *Answer*: 4.

$$E(X) = \frac{2 \cdot 6 + 3 \cdot 45 + 4 \cdot 95 + 5 \cdot 57}{203} = 4.$$

6.4.12. *Solution* (T. Cherganov). *Answer*: no.

Let X be the number of matches of the selected number. Find the mass function X:

$$\left(0, \frac{5^3}{6^3}\right), \left(1, \frac{3 \cdot 5^2}{6^3}\right), \left(2, \frac{3 \cdot 5}{6^3}\right), \left(3, \frac{1}{6^3}\right).$$

Let Y be the number of bets won. The expected value of Y is

$$E(Y) = \frac{2 \cdot 3 \cdot 5^2 + 3 \cdot 3 \cdot 5 + 4}{6^3} \approx 0.92 < 1.$$

6.4.14. The probability that a given sequence of k consecutive digits consists of the same digit is equal to

$$\frac{m(m-1)\ldots(m-k+1) + n(n-1)\ldots(n-k+1)}{(m+n)(m+n-1)\ldots(m+n-k+1)}.$$

Multiplying it by the total number of subsequences (i.e., $m+n-k+1$), we obtain the desired expected value.

6.4.15. The four aces divide the deck into five pieces. Since the average lengths of these pieces are equal, the number of cards drawn before the first ace, on average, is 48/5 (see Problem 40 in [**Mos**]).

5. Bernoulli trials (3)

Bernoulli trials are a sequence of n independent random variables, each of which takes two values: 1 with probability p and 0 with probability $q = 1-p$. Usually the appearance of 1 is called *success*, and the appearence of 0 is called *failure*.

Here is another definition of Bernoulli trials. Let M be a set of n-dimensional vectors with coordinates 0 or 1, and for each $x \in M$ the probability $P(x)$ is defined to equal $\prod_{i=1}^{n} p_i$, where $p_i = p$ if $x_i = 1$ and $p_i = q = 1 - p$ if $x_i = 0$. Elements of the set M are also called *Bernoulli trials*.

Comment. Both definitions are equivalent in the following sense. Obviously, the random variables x_i defined on the set M are independent and each of them takes value 1 with probability p and 0 with probability q. Therefore each vector $x \in M$ can be viewed as a set of values of n independent random variables x_i.

The random variable $X = \sum x_i$ is called the *number of successes*.

6.5.1. In n Bernoulli trials with probability of success p, find
   (a) the probability of exactly k successes;
   (b) the average number of successes;
   (c) the variance of the number of successes;
   (d) the most probable number of successes.

6.5.2. The law of large numbers. Let X be the number of successes in n Bernoulli trials with probability of success p. Let $t > 0$. Prove that

$$P\left(\left|\frac{X}{n} - p\right| \geq t\sqrt{\frac{pq}{n}}\right) \leq \frac{1}{t^2}.$$

Suggestion. Apply Chebyshev's inequality.

The law of large numbers means that with a large number of trials, the probability of the event "the frequency of successes is significantly different from the probability of success" is small. In fact, this law is valid not only for Bernoulli trials: if you observe a large number of independent outputs of an arbitrary random variable, then with large probability their average will be only slightly different from its expectation. This law allows, for example, a survey of a group of randomly selected people (large enough but a small part of the entire population) to be used to make conclusions about opinions and preferences.

6.5.3. The probability of having a boy is 0.515. Find the probability that among 6 children there are no more than 2 girls.

6.5.4. A factory ships iron beams. The average beam length is $3m$, with variance $0.09m^2$. How many beams must be ordered so that with a probability of at least 0.999 at least 1000 of them are at least $2m$ long?

6.5.5. Find the average number of trials before the first success if the probability of success is p.

6.5.6. Independent tests are carried out with a success probability of 0.8. Tests are carried out until the first success, but no more than four times. Find the average number of trials.

6.5.7. (Challenge) An old man was catching fish for exactly thirty-three years. Every day he caught exactly seven fish, which was just enough for dinner. The old cat living with the old man's wife eats only mackerel, which is caught half as frequently as the other fish. As a result, the cat went hungry 700 times. Do mackerel swim in the sea in schools or alone?
   Comment. Of course, a precise answer to the question posed is impossible. However, you can evaluate which of the two hypotheses fits the data better.

Suggestions, solutions, and answers

6.5.1. *Answers*: (a) $\binom{n}{k}p^k q^{n-k}$;

(b) np;

(c) npq;

(d) $\lfloor np \rfloor$ if $\{np\} \le q$, and $\lfloor np \rfloor + 1$ if $\{np\} \ge q$ (if equal, the corresponding probabilities coincide).

Suggestion. Find the ratio of probabilities of exactly k successes and exactly $k + 1$ success.

6.5.5. *Answer*: $p(1 + 2q + 3q^2 + 4q^3 + \cdots) = 1/p$.

6.5.7. If the mackerel swim in shoals, then the probability for the cat to stay hungry would be equal to $2/3$; i.e., for 33 years, the hungry days would be significantly more than 700. If the mackerel swim alone, then the probability of staying hungry is $(2/3)^7$, which is quite consistent with the data of the problem.

6. Random walks and electrical circuits[2] (3)
By A. A. Zaslavsky, M. B. Skopenkov, and A. V. Ustinov

In this section, we prove the following classical result.

Pólya's theorem. *If a person is walking randomly around a two-dimensional square grid, then he will someday return to the starting point with probability 1. If he is walking around a three-dimensional cubic grid, then the probability of return is strictly less than 1.*

All necessary definitions are given below.

The proof is based on a wonderful physical interpretation using electrical circuits. The proof itself is contained in problems 6.6.13–6.6.14, 6.6.18–6.6.24, 6.6.27–6.6.29 (with the words "theorem," "principle," and others in bold type), with the remaining problems to help develop ideas. In the process, we will get acquainted with the basics of probability theory and voltage theory. Our exposition in many ways follows [**PS**] and [**SSU**]. Another approach to the proof can be found in [**Sob**].

To solve these problems, it is useful, but not necessary, to be familiar with probability, graphs, and systems of linear equations. No special knowledge of physics is required.

One-dimensional random walk

We first formulate the problem and only then give the necessary definitions.

6.6.1. A man walks along an endless street divided into blocks (see Fig. 1 on the left). His house is at location 0 and a bar is at location 3. Starting

---

[2]The authors are grateful to I. I. Bogdanov, A. Ya. Kanel-Belov, M. V. Prasolov, A. I. Sgibnev, D. S. Chelkak, G. R. Chelnokov, F. A. Sharov, and A. Yu. Yuriev for valuable comments.

at location x between the house and the bar, he moves with probability $1/2$ one block to the left and with probability $1/2$ one block to the right. At this new location he again chooses the direction of movement at random, and everything repeats.

FIGURE 1. Random walking along the street (see problem 6.6.1) and the electrical circuit (see problem 6.6.9).

(a) Write a computer program simulating the movement of this person. Run it many times and determine the percentage of cases in which he comes home before he first reaches the bar. Use this method to guess the answers to the next problems.

(b) Let $P_T(x)$ be the probability that the person who started at location x and making no more than T moves reaches home before first reaching the bar. Fill in Table 1 with values accurate to two decimal places.

TABLE 1. Probabilities $P_T(x)$ for small T.

T \ x	0	1	2	3
1	1.00	0.50	0.00	0.00
2				
3				
4				

(c) Find the probability $P(x)$ that a person will reach home after any number of moves, but before first reaching the bar.

Definition. By a *path* of length T we mean an ordered collection of $T+1$ points of a line with integer coordinates, with adjacent points one unit apart. We say that the path *reaches* 0 *before* 3 if 0 occurs at least once in it and 3 never occurs before the first appearance of 0. The fraction of paths of length T starting at point x, reaching 0 before 3, we call the *probability $P_T(x)$* to get from x to 0 before 3 in no more than T steps. The *probability $P(x)$* to get from x to 0 before 3 is the smallest real number $P(x)$ satisfying the condition $P(x) \geq P_T(x)$ for all positive integers T. In particular, $P(0) = 1$ and $P(3) = 0$.

6.6.2. Draw all paths of length 3 that start at 1. How many are there? Which ones reach 0 before 3, and what is the proportion?

6.6.3. Pete and Paul play a game where they bet coins. Together they have 3 coins. At each turn, Pete wins from Paul one coin with probability 1/2 and loses with probability 1/2. They play until Pete has 0 coins (he lost) or 3 coins (he won all Paul's coins). Find the probability $P(x)$ that Pete will win if he starts the game with x coins.

6.6.4. (a) Solve problem 6.6.1(c) under the assumption that the bar is at the point n.

(b) A drunkard is one step away from the edge of a cliff. He steps randomly either to the edge or away from it with equal probabilities. If he reaches the edge, he falls. What are the chances of the drunkard avoiding a fall?

(c) Formulate and prove Pólya's theorem for a one-dimensional "grid."

Biased random walk*

If the probabilities of moving to the right and left are different, then the random walk is called "biased" or "asymmetrical." (The problems of this subsection are not used subsequently.)

6.6.5. Solve problems 6.6.4(a) and (b), assuming that the "traveler" moves to the right with probability p and to the left with probability $q = 1 - p$.

6.6.6. Suppose you gamble for money; initially you have 20 coins, and your opponent has 50. For each bet, you win one coin with probability 0.45 and lose with probability 0.55. The betting continues until one of the participants runs out of money. Find the probability that you go bankrupt.

6.6.7. Starting at the origin, a particle moves to the right with probability $p > 1/2$ or to the left with probability $q = 1 - p$. Find the average number of returns to the origin.

6.6.8. In a box there are n black and m white balls ($n > m$). Balls are taken out of the box one by one. Find the probability during the course of this process that there is never a time when there were equal numbers of black and white balls in the box.

Physical interpretation

A physical interpretation using electrical circuits is useful for investigating random walks. Since we are going to apply electrical circuits to prove mathematical results, we need a formal axiomatic definition.

Definition. An *electrical circuit* is a finite set of broken lines (called *wires*) on a plane or in space, possessing the following properties:

>   **isolation:** no two wires have common points, except for common end-points;
>
>   **connectivity:** any two endpoints of different wires are connected by a chain of wires

and additional structure:

- each wire is assigned a positive number (called its *conductance*);
- some wire endpoints are marked with one of the signs "+" or "−", and *each of these signs occurs.*

(In other words, an electric circuit is a finite connected graph with positive numbers placed on its edges, and at some vertices there are one of the signs "+" or "−", *with each of both signs occurring.*)

The conductance of the wire xy is denoted by $C(xy)$. The reciprocal of the conductance is called *resistance*. The endpoints of all wires are called the *vertices* of the circuit. The sets of vertices marked with the signs "+" and "−" are denoted by P and N, respectively. Informally, the meaning of these signs is: the endpoints of the wires marked with the sign "−" are connected to the negative pole of a battery (with unit voltage) and to the ground, and those with the sign "+" are connected to the positive pole; see Fig. 1 on the right.

Definition. Now let each vertex x of the electric circuit be assigned a real number $v(x)$. We call the function $v(x)$ the *voltage* if the following two axioms are satisfied.

1. *Boundary conditions.* If $x \in N$, then $v(x) = 0$. If $x \in P$, then $v(x) = 1$.

2. *Kirchhoff's current law.* If $x \notin P \cup N$, then $\sum_{xy} C(xy)(v(x) - v(y)) = 0$, where summation is over all *wires* xy containing the vertex x.

Informally, Kirchhoff's current law means conservation of charge: the sum of the currents flowing into x is equal to the sum of the currents flowing out of it. (We will give a definition of current later.)

6.6.9. Three wires of unit conductance are connected in series, as shown in Fig. 1 on the right. Find the voltages $v(x)$ at points $x = 0, \ldots, 3$.

We also give a mechanical interpretation of random walk (although we will not use it).

Definition. A *system of springs* is a finite set of line segments (called *springs*) on a plane or in space, with the following properties:

>   **isolation:** none of the springs have common points, except for common endpoints;
>
>   **connectivity:** any two endpoints of distinct springs are connected by a chain of springs

and additional structure:

- a positive number is assigned to each spring (the *stiffness*);
- some endpoints of springs are marked with one of the signs "+" or "−", *each of which occurs.*

The stiffness of spring xy is denoted by $k(xy)$. The sets of points marked with the signs "+" and "−" are denoted by P and N, respectively. Informally, the meaning of these signs is as follows: the endpoints of the springs marked with "+" and "−" are fixed at two points on the line with coordinates 1 and 0, respectively.

Definition. Now let a real number $r(x)$ be assigned to each endpoint x of the springs. We call the function $r(x)$ an *equilibrium* if the following two axioms are satisfied.

1′. *Boundary conditions.* If $x \in N$, then $r(x) = 0$. If $x \in P$, then $r(x) = 1$.

2′. *Equilibrium conditions.* If $x \notin P \cup N$, then $\sum_{xy} k(xy)(r(x) - r(y)) = 0$,

where the summation is over all *springs xy* containing the vertex x.

6.6.10. Three springs of unit stiffness are connected in series, and the endpoints of the resulting chain are fixed at points 0 and 1. Find the equilibrium values $r(x)$.

We give a mathematical definition of a random walk on a two-dimensional grid with selected *disjoint* subsets of P and N.

Definition. By a *path* of length T we mean an ordered set of $T + 1$ grid nodes, and the nodes adjacent in the set are adjacent in the grid as well. We say that the path *reaches N before P* if at least one node from the set N is encountered in it and there are no nodes from P before the first visit to a node in N. The proportion of paths of length T that start at x and reach N before P is called the *probability $P_T(x)$* of hitting N before P from x in no more than T steps. The smallest real number $P(x)$ satisfying the condition $P(x) \geq P_T(x)$ for all positive integers T is defined to be the *hitting probability $P(x)$* of the event that, starting from x, we reach N before P. In particular, $P(N) = 0$ and $P(P) = 1$.

The *probability P_T* of returning to the initial point for a random walk along a two-dimensional grid in no more than T steps is defined to be the proportion of paths of length T which return to the starting point (here $P = N = \emptyset$). The *probability* of returning to the starting point eventually is defined to be the smallest real number P satisfying the condition $P \geq P_T$ for all T.

Pólya's theorem states that the latter probability is equal to 1, but we have not proven it yet.

6.6.11. The map of a city is shown in Fig. 2 on the left. Segments indicate streets. A criminal escapes from the police. The escape routes are marked with the letter P, and the points occupied by the police are marked with the letter N. Find, with accuracy up to thousandths, the probability $P(x)$ of the event that, starting at x, the criminal will escape and will not fall into the hands of the police. From current point $x = (m, n)$ he moves to each of the points $(m+1, n)$, $(m-1, n)$, $(m, n+1)$, $(m, n-1)$ with probability $1/4$. If he reaches one of the points P or N, then his movements end.

6.6.12. Find, with accuracy up to thousandths, the voltages $v(x)$ in the circuit of wires of unit conductance in Fig. 2 on the right.

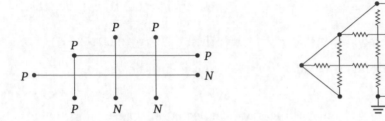

FIGURE 2. Random walk in a city and electrical circuit; see problems 6.6.11 and 6.6.12.

Existence and uniqueness of voltage

Solving the last couple of problems, we saw that the probabilities coincide with the voltages in corresponding electric circuits. Our immediate goal is to prove that this is always true.

6.6.13. (a) **Superposition principle.** If some functions $v(x)$ and $u(x)$ satisfy axiom 2, then for any real a and b, the function $av(x) + bu(x)$ also satisfies this axiom.

(b) **Maximum principle.** Each function $v(x)$, satisfying axiom 2, reaches its maximum at vertices from the set $P \cup N$; i.e., the maximum of the function $v(x)$ at all vertices of the circuit is equal to the maximum of this function on the set $P \cup N$.

(c) **Uniqueness theorem**. If $v(x)$ and $u(x)$ are two functions satisfying axioms 1 and 2, then $v(x) = u(x)$ for all x.

(d) **Existence theorem.** In any electric circuit there exists a voltage.

In subsequent problems, the definition of the hitting probability is similar to the one given above.

6.6.14. *Physical interpretation of the hitting probability.* For any electrical circuit in which all conductances are equal to 1 and such that from each vertex the same number of wires emanate (each vertex has the same number of neighbors), the voltage $v(x)$ of a vertex x is equal to the probability $P(x)$ that a random walk starting from x will hit the set P before hitting the set N.

Conductance of circuits

6.6.15. A spider moves along the edges of
(a) a cube; (b) an octahedron; (c)* a dodecahedron; (d)* an icosahedron.
It starts at a. From each vertex, it randomly (and uniformly) selects which edge to move on and travels to the next vertex. What is the probability that it will hit the opposite vertex h before it returns to the initial vertex a? (See Fig. 3 on the left.)

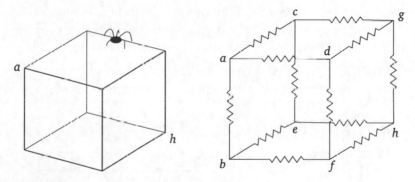

FIGURE 3. Random walks on a cube and electric circuit; see problems 6.6.15(a) and 6.6.16(a).

Of all the physical quantities associated with electric circuits, so far we have only investigated the voltages at the vertices. For Pólya's theorem, we need a few more concepts that also have probabilistic meaning.

Definition. Let $v(x)$ be the voltage in a circuit. The value $i(xy) := C(xy)(v(x) - v(y))$ is called the *current* flowing along the wire xy in the direction from x to y; $i(x) := \sum_{xy} i(xy)$ is called the *current* flowing into the circuit through the vertex x (so $i(x) = 0$ for every $x \notin P \cup N$ by axiom 2); value $C := \sum_{x \in P} i(x)$ is called the *effective circuit conductance*. The *conductance of a circuit between two sets* is the effective conductance of a circuit in which instead of P and N we use the two given sets.

For example, for the electric circuit in Fig. 1, the current through each edge is the same:

$$i(x, x+1) = v(x) - v(x+1) = (1 - x/3) - (1 - (x+1)/3) = 1/3.$$

The conductance of this circuit is equal to $C = i(0,1) = 1/3$.

6.6.16. Find the effective conductance between
(1) opposite vertices; (2) adjacent vertices of
(a) a cube; (b) an octahedron; (c)* a dodecahedron; (d)* an icosahedron with edges of unit conductance; see Fig. 3 on the right.

For practical calculations of resistance, the following problem is useful.

6.6.17. Theorem on electrical transformations. The following transformations preserve the effective conductance of a circuit:
(a) replacing two wires of conductances C_1 and C_2 connected in *parallel* (that is, having two common vertices) with one wire of conductance C_1+C_2; see Fig. 4 on the left;
(b) replacing two wires of conductances C_1 and C_2 connected in *series* (that is, having exactly one common vertex and no other wires come out of it and it does not belong to $P \cup N$) with one wire of conductance $\frac{1}{\frac{1}{C_1}+\frac{1}{C_2}}$;
see Fig. 4 on the right;
(c) combining two vertices with the same voltages into one new vertex.

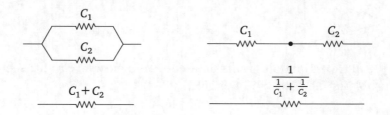

FIGURE 4. *Series and parallel* connections; see problem 6.6.17.

Now we clarify the probabilistic meaning of conductance.

6.6.18. The physical interpretation of the probability of a return. Suppose that in an electrical circuit the set N consists of one vertex (which we also denote by N) and the conductance of all wires is 1. Then the probability that a random walk along this circuit starting from the vertex N hits the set P before the first return to the vertex N is equal to $C/\deg N$, where C is the conductance of the circuit between N and P and $\deg N$ is the number of wires emanating from vertex N.

The variational principle

It seems apparent that the concept of circuit conductance is very similar to the probability of returning referred to in Pólya's theorem. We will need to investigate another property of conductance (the principle of cutting and shorting; see 6.6.22), which seems quite natural, but which is rather difficult to prove.

Denote the vertices of an electric circuit not belonging to P and N by the numbers $1, \ldots, k$, and denote the vertices belonging to P or N by numbers $k+1, \ldots, n$. Let $v(x)$ be an arbitrary function on the vertices of the circuit satisfying axiom 1. The *thermal power* of a circuit is the quantity $Q := \sum_{xy} C(xy)(v(x) - v(y))^2$, where the summation is carried out over all wires. Let $v_1 = v(1), \ldots, v_k = v(k)$. We will consider the thermal power $Q(v_1, \ldots, v_k)$ as a function of the variables $v_1, \ldots, v_k$.

6.6.19. *Reaching a minimum.* The function $Q(v_1, \ldots, v_k)$ attains a minimum value.

6.6.20. The variational principle. The function $Q(v_1, \ldots, v_k)$ attains its minimum at the point $(v_1, \ldots, v_k)$ if and only if the function $v(x)$ is a voltage.

6.6.21. The law of conservation of energy (Tellegen's theorem). The minimum value of the thermal power $Q(v_1, \ldots, v_k)$ is equal to the effective conductance of the circuit.[3]

6.6.22. The principle of cutting and shorting. Removing any *wires* of the circuit ("cutting") can only reduce the effective conductance of the circuit; see Fig. 5 in the center. Combining any vertices *(not belonging to P and N)* into one vertex ("shorting") can only increase the effective conductance of the circuit; see Fig. 5 on the right.

Two-dimensional random walk

6.6.23. (a) The conductance between the center and the boundary of the 4×4 square grid of wires of unit conductance is less than 3; see Fig. 5 on the left.

(b) **The conductance of a square.** What does the conductance between the center and the boundary of the $2n \times 2n$ square grid (of wires of unit conductance) tend to when n tends to infinity?

---

[3] It would be more correct to say "effective conductance times the square of the battery voltage" but we have a battery voltage equal to 1.

(c)* Find the *order of decrease* of the conductance as a function of n; *i.e.,*
*find a function $f(n)$ such that the conductance is bounded between $\frac{1}{100}f(n)$
and $100f(n)$.*

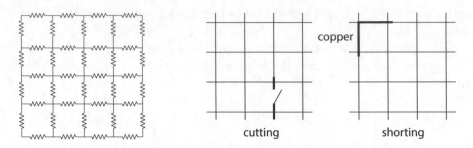

FIGURE 5. Square grid 4×4, cutting and shorting; see problems 6.6.23 and 6.6.22.

6.6.24. Prove Pólya's theorem for two-dimensional random walks.

Three-dimensional random walks

In proving Pólya's theorem in two dimensions, we combined together
("shorted") *vertices* of the electrical circuit to evaluate a lower bound for
the probability of a return. For the three-dimensional case, we will need to
estimate an upper bound for the probability of a return. Consequently, we
will instead cut some of the *wires* from our circuit. We will strive to obtain
a circuit whose conductance is easy to calculate. Trees are ideal for this.

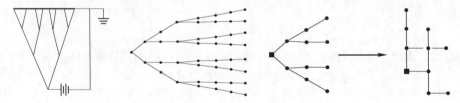

FIGURE 6. (Left) a binary tree of depth 3; (in the center) a
modified binary tree of depth 3; (right) permissible intersections of edges in this tree; see problems 6.6.25, 6.6.27, and
6.6.28.

6.6.25. Find the resistance of a binary tree with depth
    (a) 3;
    (b) 2010, composed of *wires with unit conductance* (see Fig. 6 on the
left).

6.6.26. Which of the trees mentioned in problem 6.6.25 can be cut out from
a two-dimensional grid?

6.6.27. *Resistance of a tree.* Find the resistance of the *modified* binary and ternary trees of depth n, in which every wire at the kth level is replaced by 2^k wires with unit conductance that are connected in series (see Fig. 6 in the center).

6.6.28. *Pruning a tree.* Is it possible to cut wires from a two-dimensional grid to produce the modified binary tree of problem 6.6.27 if we are allowed to join some vertices (but not *wires*) that are located an equal distance from the root? (See Fig. 6 on the right.) The same question, but now we ask about a ternary tree, extracted by cutting from the three-dimensional grid.

6.6.29. *Three-dimensional case.* Prove Pólya's theorem for a three-dimensional grid.

Suggestions, solutions, and answers

6.6.1. (a) The correctness of the program can be estimated as follows: the difference between the "real" and calculated probabilities should be approximately proportional to the number $\frac{1}{\sqrt{n}}$, where n is the number of experiments.

(b) *Answer:* see Table 2.

Suggestion (P. Belopashentseva). Fill in the table, either directly listing the paths and calculating the corresponding probabilities or by using the equations $P_T(0) = 1$, $P_T(3) = 0$, $P_T(x) = \frac{1}{2}(P_{T-1}(x-1) + P_{T-1}(x+1))$ for $x = 1, 2$.

TABLE 2. Probabilities $P_T(x)$ and $P(x)$.

T \ x	0	1	2	3
1	1.00	0.50	0.00	0.00
2	1.00	0.50	0.25	0.00
3	1.00	0.63	0.25	0.00
4	1.00	0.63	0.31	0.00
$P(x)$	1.00	0.67	0.33	0.00

(c) *Answer:* $P(x) = 1 - x/3$; see the last row in Table 2. This problem is a special case of problem 6.6.4(a).

6.6.3. *Answer:* $P(x) = x/3$; this problem is equivalent to problem 6.6.1(c).

6.6.4. (a) *Answer:* $P(x) = 1 - \frac{x}{n}$.

Solution (E. Pavlov). Let $N_T(x)$ be the set of paths of length T starting at x that reach 0 before they reach n. Then the probability $P_T(x)$, that a person who started at x and made no more than T steps will reach home before reaching the bar by definition is $P_T(x) = |N_T(x)|/2^T$, where 2^T is the total number of paths of length T. The probability $P(x)$ that a person who started at x will reach home before reaching the bar is by definition the smallest real number $P(x)$ that satisfies the condition $P(x) \geq P_T(x)$ for each T.

We verify that $P(x)$ has the following two properties:

1. $P(0) = 1$ and $P(n) = 0$;
2. $P(x) = \frac{1}{2}P(x-1) + \frac{1}{2}P(x+1)$ for each $x = 1, 2, \ldots, n-1$.

Property 1 is obvious. Let us prove property 2. To any path of length $T + 1$ starting at x we associate a path of length T obtained by removing the first step. Such a correspondence maps one-to-one $N_{T+1}(x)$ onto $N_T(x-1) \cup N_T(x+1)$ for every x, $0 < x < n$. Therefore, $P_{T+1}(x) = \frac{1}{2}(P_T(x-1) + P_T(x+1))$ for $0 < x < n$.

On the one hand, for every T, the right-hand side does not exceed $\frac{1}{2}(P(x-1) + P(x+1))$, which means that $P(x) \leq \frac{1}{2}(P(x-1) + P(x+1))$.

On the other hand, due to the minimality of $P(x-1)$ and $P(x+1)$ for each $\varepsilon > 0$ there is T such that $P_T(x-1) + P_T(x+1) > P(x-1) + P(x+1) - \varepsilon$. Therefore, $P(x) \geq \frac{1}{2}(P(x-1) + P(x+1))$. Property 2 is proved.

From properties 1 and 2 it follows that $P(x)$ is an arithmetic progression: $P(x) = 1 - \frac{x}{n}$.

Comment. It can be shown that $P(x) = \lim_{T \to \infty} P_T(x)$ (although this is not required to prove Pólya's theorem). Indeed, observe that $P_T(x) \leq 1$. Furthermore, $P_T \leq P_{T+1}(x)$, since $|N_{T+1}(x)| \geq 2|N_T(x)|$, because each path from $N_T(x)$ extends to two paths from $N_{T+1}(x)$. Thus, $P_0(x), P_1(x), P_2(x), \ldots$ is a bounded monotone sequence. Therefore it has a limit, which is equal to the exact upper bound $P(x)$ of the sequence.

(b) *Answer:* 0.

Solution (A. Balakin). Let the drunkard stand at a point with coordinate 1, and let the cliff be at 0. Let P_T be the probability of getting to point 0 during the first T steps, and let $P_{T,n}$ be the probability of getting to point 0 before reaching point n. Note that $P_{T,n} \leq P_T \leq 1$, since all paths reaching 0 before n reach 0. By (a), we have $\lim_{T \to \infty} P_{T,n} = 1 - \frac{1}{n}$. This means that for the probability $P = \lim_{T \to \infty} P_T$ to fall off the cliff, the inequalities $1 - \frac{1}{n} \leq P \leq 1$ must hold for all positive integers n. But in this case $P = 1$. Then the desired probability of avoiding a fall is $1 - P = 0$.

(c) **Theorem.** *For the random walk on the one-dimensional "grid", the probability of returning to the starting point is* 1.

Proof (from [**SSU**, § 3.1]). Let P be the probability of returning to the starting point. Denote by P_n the probability of returning to the starting point before reaching the point n or $-n$. Then $P_n \leq P \leq 1$ for any n.

Now we prove that $P_n = 1 - 1/n$. After the first move, the traveler reaches one of the points 1 and -1 with probability $1/2$. If he reaches the point 1, then by (a) we see that the probability of returning to the origin before reaching the point n is $1 - 1/n$. If he finds himself at the point -1, we reason in a like manner. Similarly to the proof of property 2 in the solution of (a), we obtain

$$P_n = \frac{1}{2}\left(1 - \frac{1}{n}\right) + \frac{1}{2}\left(1 - \frac{1}{n}\right) = 1 - \frac{1}{n}.$$

Since $1 - 1/n \leq P \leq 1$ for every n, we see that P is equal to 1.

Comment. Along the way, we *made sure* that $P = \lim\limits_{n \to \infty} P_n$, which is not so easy to establish directly. Another approach to the proof of the theorem can be found in [**Sob**].

6.6.5. *Answer:* the probability of reaching home is $P(x) = \frac{(p/q)^{3-x}-1}{(p/q)^3-1}$; the probability of falling is $P = 1$ for $p \leq 1/2$ and $P = \frac{1-p}{p}$ for $p > 1/2$.

Suggestion. To calculate the probability of reaching home, we reason the same way as in problem 6.6.4(a). Show that properties 1–2 must be replaced with the following.

$1'$. The equalities $P(0) = 1$ and $P(n) = 0$ are satisfied.

$2'$. For each $x = 1, 2, \ldots, n-1$, the equality $P(x) = qP(x-1)+pP(x+1)$ holds.

Choose A and B such that the function $f(x) = A(q/p)^x + B$ satisfies properties $1'$–$2'$.

The probability x of a drunkard falling satisfies the equation $x = 1-p+px^2$. For $p \leq 1/2$, this equation has the single root equal to 1 on the segment $[0,1]$, and for $p > 1/2$ there are two roots: $x_1 = 1$ and $x_2 = \frac{1-p}{p}$. For $p = 1$, the probability is 0. Since the probability of falling is a continuous function of p, it is equal to 1 for $p \leq 1/2$ and $\frac{1-p}{p}$ for $p > 1/2$.

6.6.6. *Answer:* $1 - \frac{(0.55/0.45)^{20}-1}{(0.55/0.45)^{70}-1}$. It is approximately 99.995%.

Suggestion. Use the suggestion to problem 6.6.5.

6.6.7. *Answer:* $\frac{1}{2(1-p)} - 1$.

Suggestion. If the first step is made to the left, then the probability of a return is equal to 1, and if it's made to the right, then it is $\frac{1-p}{p}$. Therefore, the total probability of return is $2(1-p)$, and the average number of returns equals $\frac{1}{2(1-p)} - 1$.

6.6.8. *Answer:* $\frac{n-m}{m+n}$.

Suggestion. If the first drawn ball is white, then the probability that at some moment the numbers of black and white balls taken out will be equal is 1. If the first drawn ball is black and at some point the numbers were equal, then you can change the color of all balls drawn before the first occurrence of equality to the opposite. It follows that the probability of achieving equality is $\frac{2m}{m+n}$.

6.6.9. *Answer:* $v(x) = 1 - x/3$.

Suggestion. It follows from axioms 1 and 2 that the function $v(x)$ will be linear for this circuit.

6.6.10. *Answer:* $r(x) = 1 - x/3$. This problem is equivalent to problem 6.6.9.

6.6.11. *Answer*: see Fig. 7 on the left.

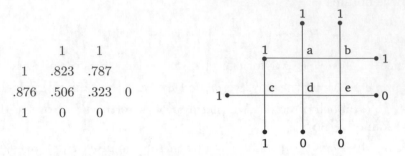

	1	1	
1	.823	.787	
.876	.506	.323	0
1	0	0	

FIGURE 7. Probabilities $P(x)$ or voltages $v(x)$; see problems 6.6.11 and 6.6.12.

Suggestion. The city map is shown in Fig. 7 on the right. The probabilities $P(x)$ are denoted by a, b, c, d, and e. As in the one-dimensional case, the function $P(x)$ satisfies axioms 1 and 2 from the definition of an electric circuit. This gives rise to a system of linear equations, whose solution is the answer to the problem:

$$a = (b + d + 2)/4;$$
$$b = (a + e + 2)/4;$$
$$c = (d + 3)/4;$$
$$d = (a + c + e)/4;$$
$$e = (b + d)/4.$$

6.6.12. *Answer*: see Fig. 7 on the left.
Suggestion. This problem is equivalent to 6.6.11.
6.6.13. (a) *Solution* (from article [**SSU**, § 2.1]). For any vertex $x \notin P \cup N$,

$$\sum_{xy} C(xy)\left(au(x) + bv(x) - au(y) - bv(y)\right)$$

$$= a\sum_{xy} C(xy)\left(u(x) - u(y)\right) + b\sum_{xy} C(xy)\left(v(x) - v(y)\right) = 0,$$

where the summation is over all the wires xy coming out of x. Therefore, $au(x) + bv(x)$ satisfies axiom 2.

(b) *Solution* (from [**SSU**, § 2.1]). Let $v(x)$ attain its maximum value at some vertex x. Let us prove that if $x \notin P \cup N$, then our function has the same value at the neighboring vertices. Since $v(x)$ is the maximum value, then for each of the neighboring vertices y, the inequality $v(x) - v(y) \geq 0$ holds. Therefore, $\sum_{xy} C(xy)(v(x) - v(y)) \geq 0$. By axiom 2, the last inequality is an equality. Therefore, $v(y) = v(x)$ for all y.

Since the electric circuit is a connected graph, there is a path, connecting the vertex x with one of the boundary vertices. By the previous argument, the values of our function at all vertices of the path will be equal. Therefore, at one of the boundary vertices, the maximum value of the function $v(x)$ is also attained.

(c) *Solution* (from [**SSU**, §2.1]). Consider the function $u(x) - v(x)$. It takes the value 0 on $P \cup N$ and, by the principle of superposition (part (a)), satisfies axiom 2. By the maximum principle (part (b)), $u(x) - v(x) \leq 0$ for all vertices x. Similarly, $u(x) - v(x) \geq 0$ for all vertices x. Hence, this function is equal to 0 at all vertices. Therefore, $u(x)$ and $v(x)$ are equal.

(d) *Suggestion*. Consider the special case when all wires have a conductance of 1 (the case of arbitrary conductance is similar). Consider a random walk on an electrical circuit. Let $P(x)$ be the probability that, starting from x, we hit the set P before the set N. The function $P(x)$ satisfies axioms 1 and 2.

Other approaches to the proof are discussed in [**SSU**, §2.2].

6.6.14. *Suggestion*. As with the one-dimensional case, the probability $P(x)$ satisfies axioms 1 and 2. The voltage $v(x)$ also satisfies these axioms. By the uniqueness theorem of 6.6.13(c), $v(x) = P(x)$ for all x.

6.6.15. *Answers*: (a) 2/5; (b) 1/2; (c) 2/7; (d) 2/5.

Suggestion. Reason similarly as in the solution to 6.6.11, but it's easier to reduce the solution to problem 6.6.16 with the help of 6.6.18.

6.6.16. *Answers*: (1)(a) 6/5; (1)(b) 2; (1)(c) 6/7; (1)(d) 2; (2)(a) 12/7; (2)(b) 12/5; (2)(c) 30/19; (2)(d) 30/11.

Suggestion. It is possible to reason similarly to the solution to problem 6.6.11, but it is easier to sequentially simplify the circuit using transformations from 6.6.17; see [**PS**, Fig. 23–25].

A particularly elegant solution to items (2)(a)–(2)(d) is obtained from the following theorem (see [**F**]) and [**SSU**, §3, Theorems 3.2 and 3.5]):

Foster's average resistance theorem. *In any electrical circuit with n vertices, the equality*

$$\sum_{xy} \frac{C(xy)}{C(x \leftrightarrow y)} = n - 1$$

holds, where the sum is taken over all the wires xy of the electric circuit and $C(x \leftrightarrow y)$ denotes the effective conductance between the vertices x and y.

6.6.17. *Solution* (from [**SSU**, §2.3]). We prove that these transformations do not change the voltages of the vertices. Let the vertices 1, 2 with voltages v_1, v_2 be connected by parallel wires with conductance C_1 and C_2. Before the transformation, the current from vertex 1 to vertex 2 is $C_1(v_2 - v_1) + C_2(v_2 - v_1)$. This equals $(C_1 + C_2)(v_2 - v_1)$, the current between the vertices 1 and 2 after the transformation, while maintaining the same voltage values. Thus axioms 1 and 2 will remain satisfied, so the conductance of the circuit will not change.

(b) *Solution* (from [**SSU**, §2.3]). We prove that during our transformation, the voltages of the vertices do not change (with the exception of the common vertex for the two wires under consideration, which disappears).

Indeed, let the vertices 1, 2, 3 with voltages v_1, v_2, v_3 be connected by two wires in series with conductance C_1 and C_2. Replace C_1 and C_2 with a single wire of conductance $\frac{C_1 C_2}{C_1+C_2}$ between 1 and 3. We intentionally leave at each vertex the previous value of the voltage.

We verify axioms 1 and 2 for the values we have chosen. It is clear that axiom 1 is still satisfied. It is also clear that axiom 2 for each of the vertices, except for 1 and 3, is also satisfied, since our transformation does not affect them. Find the currents flowing into the vertex 3. In the original circuit, the current from 2 to 3 is $C_2(v_2-v_3)$. In the new circuit, the current from 1 to 3 is equal to $\frac{C_1 C_2}{C_1+C_2}(v_1-v_3)$. But by axiom 2 for the vertex 2 of the original circuit we get $C_1(v_1 - v_2) = C_2(v_2 - v_3)$, hence $C_1(v_1 - v_3) = (C_1 + C_2)(v_2 - v_3)$, which means $\frac{C_1 C_2}{C_1+C_2}(v_1 - v_3) = C_2(v_2-v_3)$. We see that the currents flowing into the vertex 3 have not changed. Therefore, axiom 2 for the vertex 3 is still satisfied, and likewise for the vertex 1.

So, the intentionally chosen values of the voltage satisfy axioms 1 and 2 after the transformation. By the uniqueness theorem (see 6.6.13(c)) this is the voltage in the new circuit. Since the voltages of the vertices have not changed, neither the currents through the wires nor the conductance of the circuit have changed.

(c) *Solution* (from [**SSU**, § 2.3]). Let the voltages of vertices 1 and 2 be equal. Connect these two vertices without changing voltage.

Obviously, axiom 1 is still satisfied. For the new vertex (if it does not belong to $P \cup N$) obtained by combining vertices 1 and 2, axiom 2 is satisfied, since it is obtained by adding axiom 2 for the vertices 1 and 2 of the original circuit. The remaining vertices of our transformation are not affected; therefore, for them axiom 2 also remains true. Since the voltages of the vertices have not changed, then neither the currents through the wires nor the conductance of the circuit changed.

6.6.18. *Solution* (from article [**SSU**, § 2.3]). Let wires from vertex N lead to the vertices with numbers $1, 2, \ldots, k = \deg N$. Let $v(1), v(2), \ldots, v(k)$ denote their voltages. According to the physical interpretation of the probability of hitting a boundary (problem 6.6.14), the voltage $v(i)$ is equal to the probability of hitting the set P from i before reaching N.

The probabilities of taking the first step into any of the vertices $1, 2, \ldots, k$ are the same and equal to $\frac{1}{k}$. Therefore, the probability of hitting the set P, starting from N, before first reaching N is

$$\frac{1}{k}v(1) + \frac{1}{k}v(2) + \cdots + \frac{1}{k}v(k) = \frac{-i(N)}{k} = \frac{C}{k} = \frac{C}{\deg N}.$$

6.6.19. *Solution* (from article [**SSU**, § 2.4]). Denote $Y := Q(0, \ldots, 0)$. Let us prove that if at least one of the numbers $v_1, \ldots, v_k$ exceeds in absolute value the number $X := n\sqrt{Y/\min C(xy)}$, then $|Q(v_1, \ldots, v_k)| > Y$. Suppose that $|v_1| > X$. Take the shortest path connecting vertex 1 to one of the vertices of the set N. There are no more than n wires in this path, so there is a wire xy, on which $|v(x) - v(y)| > \sqrt{Y/\min C(xy)}$. On this wire the expression

$C(xy)(v(x) - v(y))^2$ already exceeds Y. Since all the summands in sum $\sum_{xy} C(xy)(v(x) - v(y))^2$ are nonnegative, we obtain the required inequality.

We restrict the domain of the function $Q(v_1, \ldots, v_k)$ to the set $[-X, X]^k$. Since the set under consideration is compact and the function $Q(v_1, \ldots, v_k)$ is continuous, the function attains its minimum value in this set. This minimal value is obviously no more than $Y = Q(0, \ldots, v_k)$, so it is not greater than the values of the function $Q(v_1, \ldots, v_k)$ outside the set $[-X, X]^k$. Therefore, this value is the minimum value for all points in $\mathbb{R}^k$.

6.6.20. *Solution* (from article [**SSU**, § 2.4]). Consider an arbitrary vertex $x \notin P \cup N$. Let, without loss of generality, it be numbered 1. Consider $Q(v_1, \ldots, v_k)$ as a quadratic function of v_1. The leading coefficient is positive, which means the smallest value is attained when $v_1 = \sum_{xy} C(xy)v(y) / \sum_{xy} C(xy)$, where the summation is over all edges xy coming out of x. So the smallest value is attained exactly when axiom 2 holds for vertex x.

6.6.21. *Solution* (from article [**SSU**, § 2.4]). We have

$$\sum_{xy} C(xy)(v(x) - v(y))^2$$

$$= \sum_{xy} (v(x)C(xy)(v(x) - v(y)) + v(y)C(xy)(v(y) - v(x)))$$

$$= \sum_{x-1}^{n} \left(v(x) \sum_{xy} C(xy)(v(x) - v(y)) \right).$$

The value $v(x) \sum_{xy} C(xy)(v(x) - v(y))$ is equal to zero for all vertices not belonging to $P \cup N$ by axiom 2 and for all vertices from set N by axiom 1. Hence,

$$Q(v_1, \ldots, v_k) = \sum_{x \in P} \sum_{xy} C(xy)(1 - v(y)) = C.$$

6.6.22. *Solution* (from article [**SSU**, § 2.4]). Let C be the conductance of the original circuit, and let C' be the conductance of the circuit after removing one of the wires. Let $Q(v_1, \ldots, v_k)$ denote the thermal power of the original circuit, and let $Q'(v_1, \ldots, v_k)$ be the new one (as a function of the variables $v_1, \ldots, v_k$). Denote by $v(1), \ldots, v(k)$ the voltages of the vertices in the original circuit, and denote by $v'(1), \ldots, v'(k)$ the voltages in the new one. Then, according to the law of conservation of energy and the variational principle (problems 6.6.21 and 6.6.20), we get

$$C = Q(v(1), \ldots, v(k)) \geq Q'(v(1), \ldots, v(k)) \geq Q'(v'(1), \ldots, v'(k)) = C'.$$

The statement about combining vertices is proved similarly.

6.6.23. (b) *Answer*: $C \to 0$ as $n \to \infty$.

Suggestion. We apply the principle of cutting and shorting (see 6.6.22): we combine together the vertices located on the boundaries of the concentric squares 2×2, 4×4, $\ldots$, $2n \times 2n$, *as shown in Fig. 8. The resulting circuit has*

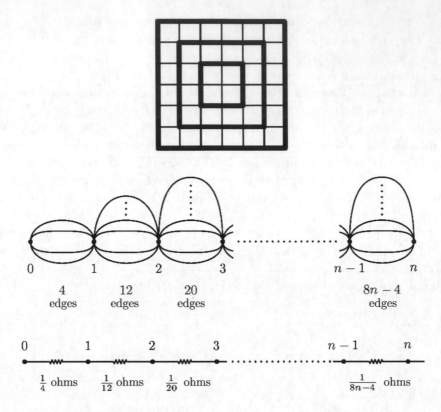

FIGURE 8. Combining in a square circuit; see solution to problem 6.6.23(b).

the same conductance as the circuit in Fig. 8 in the center. The conductance of the obtained circuit is calculated using problem 6.6.17. Since it is possible to replace i parallel wires of conductance 1 with one wire of conductance i, the circuit has the same conductance as the circuit in Fig. 8. Its conductance is equal to $\dfrac{1}{\sum\limits_{k=1}^{n} \frac{1}{8k-4}}$. *This number tends to zero as n tends to infinity. Since*

the conductance of the initial circuit is no more than the conductance of the resulting circuit, it also tends to zero.

(c) *Answer*: $f(n) = 1/\ln n$.

Suggestion. Use problem 6.6.27. From the $2n \times 2n$ square cut out a modified binary tree with a root in the center of the square (as in Fig. 10); at the same time, vertices can be joined at equal distances from the root).

Comment. To prove Pólya's theorem, it suffices to find the limit of conductance of a square (part (b)), rather than the order of its decreasing (part (c)).

6.6.24. *Solution* (from article [**SSU**, §3.2]). Let P be the probability that a random walk on a two-dimensional grid returns to the starting point. We denote by P_n the probability that a random walk returns to the starting

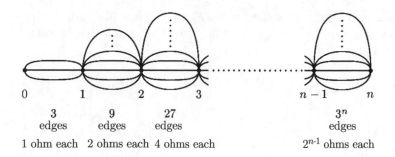

FIGURE 9. Calculation of the conductance of a tree; see the solution of problem 6.6.27.

point before reaching the boundary points of the $2n \times 2n$ square with center at the starting point.

It is clear that $P_n \leq P \leq 1$ for each n. From the physical interpretation of the return probability (problem 6.6.18) we obtain that $P_n = 1 - C/4$, where C is the effective conductance between the center and the boundary of the $2n \times 2n$ square. From the statement of problem 6.6.23(b) it follows that C tends to zero when n tends to infinity. Therefore, $P_n \to 1$ for $n \to \infty$ and P equals 1.

Suggestion to another solution. It can be proved that in a two-dimensional random walk, the average number of returns to the starting point is infinite, and for a three-dimensional walk it is finite. From this it follows that the probability of a return for the two-dimensional walk is equal to 1, and for the three-dimensional walk it is strictly less than 1.

6.6.25. Prove by induction that the resistance of a binary tree of depth n of wires with conductance equal to 1 is $1 - \frac{1}{2^n}$.

6.6.26. It is easy to cut out a binary tree with depth 3. We show that a binary tree with depth 2010 cannot be cut. If you managed to cut it, then all its vertices are located at a distance of no more than 2010 from the root; from here we get that the tree is in a square with side $2 \cdot 2010 + 1$. Therefore, the number of its vertices does not exceed $4021^2 \leq 2^{24}$. On the other hand, the number of its vertices is $2^{2011} - 1$. This contradiction completes the proof.

6.6.27. *Answer* (for ternary tree): $1 - \frac{2^n}{3^n}$.

Suggestion (see [**SSU**, § 3.3, the lemma about conductance of a tree]). The voltages at points located at the same distance from the root of the tree are equal by symmetry. Join these points. We get the circuit shown in Fig. 9. The circuit resistance has not changed and it is equal to

$$\frac{1}{3} + \frac{2}{9} + \cdots + \frac{2^{n-1}}{3^n} = 1 - \frac{2^n}{3^n}.$$

6.6.28. Cutting a modified binary tree from a plane and a ternary tree from the space in a similar way (Fig. 10). The proof is by induction on the depth of a tree.

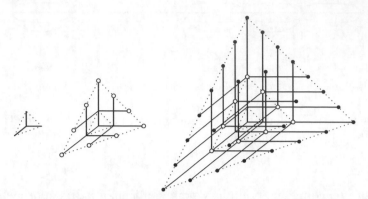

FIGURE 10. Cutting out a ternery tree with intersections in space; see solution to problem 6.6.28.

6.6.29. *Solution* (A. Yuriev) (see [**Yu**] and [**SSU**, § 3.3]). We prove that the probability of a return to the starting point does not exceed 5/6.

Take an arbitrary $i > 0$, put $n = 2i - 1$, and consider a part of the grid inside the octahedron $\{(x, y, z) : |x| + |y| + |z| \le n\}$ with a diagonal of $2n$. Let C_n be the conductance between the origin and the boundary of the octahedron. By problem 6.6.28, a modified ternary tree of depth i can be cut from this part of the grid with the intersections of wires at an equal distance from the root. By problem 6.6.17(c), the conductance of a tree with such intersections is equal to the conductance of the same tree without intersections. By problem 6.6.27, the conductance of modified ternary trees of any depth is greater than 1. The principle of cutting and shorting shows that $C_n > 1$.

By the physical interpretation of the probability of return, it follows that the probability of a return to the center of the octahedron before reaching its boundary points is $P_n = 1 - C_n/6 < 5/6$.

Let P_n^T be the probability of returning to the center of the octahedron with a diagonal of $2n$, without reaching its border and making no more than T steps. Let P^T be the probability of returning to the starting point, traveling in the three-dimensional grid, and making no more than T steps. For $T < n$, the octahedron boundary cannot be reached in T steps. It follows that $P^T = P_n^T$ for $T < n$, since under such restrictions on T the walk on the octahedron does not differ from the walk on the three-dimensional grid. Using the fact that $P_n^T \le P_n$, we arrive at the estimate $P^T \le P_n < 5/6$ for $T < n$.

Therefore, for every T, we have $P^T < 5/6$ and then by definition the probability of returning to the starting point, traveling in the three-dimensional grid is also less than 5/6.

Chapter 7

Combinatorial geometry

1. Rug runners and napkins (2)
By P. A. Kozhevnikov

This collection of problems gives you a sense of what combinatorial geometry is about. It will be helpful to have familiarity with the extremal principle and the pigeonhole principle (Sections 1 and 2 of Chapter 2).

One-dimensional geometry, or "rug runners"

In problems 7.1.1–7.1.6 it is assumed that several "rug runners" (segments, since they are one-dimensional) are laid in a long narrow corridor (on a straight line).

7.1.1. It is known that any two rug runners intersect. Prove that you can nail all these rug runners to the floor with one nail.

7.1.2. It is known that each rug runner intersects at least with half of the others. Prove that there exists a rug runner that intersects all the others.

7.1.3. It is known that each point of the corridor is covered by no more than k rug runners. Prove that the set of all rug runners can be divided into k subsets so that in each subset the rug runners do not intersect.

7.1.4. It is known that all rug runners have the same length. Prove that you can drive several nails into the floor so that each rug runner is nailed with exactly one nail.

7.1.5. It is known that the corridor is l units long and the rug runners completely cover it.

(a) Prove that you can remove some rug runners so that the remaining rug runners covered the entire corridor and their total length does not exceed $2l$.

(b) Prove that some rug runners can be removed so that the remaining rug runners do not intersect and their total length is not less than $l/2$.

7.1.6. It is known that the corridor is l units long and the rug runners completely cover it. Each rug runner was cut in half and one of the halves was removed. Prove that the remaining halves will cover at least one third of the length of the corridor. (*D. Fomin*, Leningrad Mathematical Olympiad 1990).

7.1.7. Each of seven children skated three times on Sunday. It is known that every two of them were on the ice at the same time at some moment. Prove that, in fact, three children were on the ice at the same time at some moment. (*A. Anjans*, All-Union Mathematical Olympiad 1989)

Two-dimensional geometry, or "napkins on the table"

In problems 7.1.8–7.1.13 it is assumed that several rectangular napkins are put on a rectangular table in such a way that their sides are parallel to the edges of the overlap.

7.1.8. It is known that every two napkins overlap. Prove that you can anchor all the napkins to the table with a single nail.

7.1.9. It is known that each napkin overlaps with at least 3/4 of the remaining ones. Prove that some napkin overlaps with all other napkins.

7.1.10. Let napkins be equal squares arranged in such a way that each point of the table is covered by at most k napkins. Prove that the napkins can be partitioned into $2k - 1$ groups so that no two napkins in the same group overlap.

7.1.11. Let the napkins be equal squares. Prove that you can hammer in several nails so that each napkin is nailed with exactly one nail. (*A. Berzins and I. Izmestiev*, All-Russian Mathematical Olympiad 1995)

7.1.12. Let napkins be unit squares arranged so that the distance between the centers of any two napkins does not exceed 2. Prove that you can place one more such napkin on the table that will overlap with all the napkins already on the table. (*A. Anjans*, Tournament of Towns, 1990)

7.1.13. On a 20×20 table there are 90 unit square napkins. Prove that you can place one more napkin on the table that does not overlap with any of those aleady on the table.

In problems 7.1.14 and 7.1.15, the napkins and the table are rectangular, but the sides of the napkins are not necessarily parallel to the edges of the table.

7.1.14. Let the size of the table be 20×20, let each napkin be of size $a \times 1$, and let there be a total of 95 napkins on the table. Prove that you can draw a circle of diameter 1 on the table that does not overlap with any napkin.

7.1.15. The total perimeter of the napkins on a 20×20 table is greater than 200. Prove that you can draw a straight line on the table that intersects at least 6 napkins.

Three dimensions

7.1.16.* Try to formulate three-dimensional analogues of the problems above.

7.1.17. Twelve rectangular parallelepipeds are placed in space, with edges parallel to the coordinate axes. Construct a graph whose vertices correspond to the parallelepipeds, where we connect two vertices of the graph with an edge if and only if the corresponding parallelepipeds do not intersect. Could this graph be a cycle of length 12? (*A. Akopyan*, All-Russian Mathematical Olympiad 2005)

7.1.18. Several convex polyhedra are arranged in space so that any two of them intersect. Prove that there is a plane which intersects all of them.

7.1.19. "Suitcase in the subway". Is it possible to place a rectangular box with the sum of its edges greater than d into a rectangular parallelepiped with the sum of its edges equal to d (*A. Shen*, Tournament of Towns, 1998)?

7.1.20. The same question for tetrahedrons.

Suggestions, solutions, and answers

When solving these problems, it is helpful to use the *extremal principle*. The idea is to consider an object which is distinguished among all other objects by some special property. Initial arguments using the extremal principle might be, for example, as follows: consider the leftmost endpoint of a set of line segments, or the pair of most distant points of a given set, etc.

7.1.1. Let $[a_i, b_i]$, $i = 1, 2, \ldots, n$, be the given segments. Let $a = a_i = \max\{a_1, a_2, \ldots, a_n\}$, $b = b_j = \min\{b_1, b_2, \ldots, b_n\}$. If $a > b$, then the segments $[a_i, b_i]$ and $[a_j, b_j]$ do not overlap, which is a contradiction. Therefore, $a \leq b$, and all segments contain the segment $[a, b]$.

Comment. This problem is Helly's theorem in the one-dimensional case (see Section 2 of this chapter). It remains true for an infinite number of segments. The proof for an infinite set is similar, but instead of the largest and smallest numbers, we consider the infimum and supremum.

7.1.2. Let $[a_i, b_i]$, $i = 1, 2, \ldots, n$, be the given segments, $a = a_i = \max\{a_1, a_2, \ldots, a_n\}$, $b = b_j = \min\{b_1, b_2, \ldots, b_n\}$. If $a \leq b$, then all segments contain $[a, b]$. Otherwise, more than $n/2$ segments intersect $[a_i, b_i]$ and thus contain the point $a = a_i$ (we assume that an arbitrary segment intersects itself). Also, more than $n/2$ segments intersect $[a_j, b_j]$ and, therefore, they contain the point $b = b_j$. By the pigeonhole principle, there is a segment containing both a and b. This segment intersects all other segments.

7.1.3. Let $[a_i, b_i]$, $i = 1, 2, \ldots, n$, be the given segments. We apply induction on n (the base is obvious). Let $a = a_i = \max\{a_1, a_2, \ldots, a_n\}$. Any segment intersecting $[a_i, b_i]$ covers a_i; thus $[a_i, b_i]$ intersects at most $k - 1$ segments. Having deleted the segment $[a_i, b_i]$, we apply the induction hypothesis to the remaining segments. Having returned $[a_i, b_i]$, we assign it to the set, which does not include segments intersecting with it.

Comment. This argument can be used to solve the following problem from the Moscow Olympiads: *Fifty segments lie on a line. Prove that either there are 8 segments having a common point or there are 8 pairwise disjoint segments.*

7.1.4. Introduce coordinates on the line in such a way that all segments have a unit length and no endpoint of a segment has an integer coordinate. Now it is enough to hammer nails into integer points.

7.1.5. (a) Remove a segment if it is covered by the union of other segments. It is not difficult to prove that after a finite number of such steps we are left with a collection of segments such that each point is covered by at most two segments.

(b) It can be noticed that the covering obtained after completing the steps of part (a) is the union of two sets of segments, such that in each of them segments are pairwise disjoint.

7.1.6. Let the segments be $\Delta_1, \ldots, \Delta_k$, after removing one half of each of the original segments. We "inflate" Δ_i by a factor of three; i.e., we consider the segment Δ_i' obtained from Δ_i by a homothety with center in the midpoint and coefficient 3. It is easy to show that $\Delta_1', \ldots, \Delta_k'$ cover the corridor.

7.1.7. The time spent at the skating rink by each skater is a set A_i of three pairwise disjoint segments on the time axis t. Their union is the set

$$E = \bigcup_{i=1}^{7} A_i \text{ of } 3 \cdot 7 = 21 \text{ line segments. The set } V \text{ of the endpoints of all these}$$

segments consists of 42 points (the endpoints can be considered distinct). Estimate the number of pairs (e, v) for which the segment $e \in E$ contains the endpoint $v \in V$ of a segment other than e. Each intersection $A_i \cap A_j$ produces at least two such pairs. Therefore there are at least $2\binom{7}{3} = 42$ such pairs. In addition, the leftmost endpoint of the segments in E does not belong to any such pair. Therefore, some endpoint of a segment belongs to at least two pairs, i.e., is covered with two other segments. This endpoint of this segment corresponds to the time moment we are looking for.

For two- and three-dimensional problems, the idea of *reducing the dimension* is useful. For example, by projecting a configuration onto straight lines (say, on the l- and m-axes parallel to the table edges), we can sometimes reduce the problem to simpler one-dimensional problems.

7.1.8. *Project* the rectangular napkins onto the lines l and m parallel to the edges of the table. By problem 7.1.1, the projections onto l have a common point A. This means that the line m' parallel to m and passing through A intersects all the rectangles. Similarly, there is a line l' parallel to l intersecting all the rectangles. It is easy to see that the point $l' \cap m'$ belongs to all the rectangles.

7.1.9. *Project* n of the rectangles onto l and m, as above. Following the solution to problem 7.1.2, we prove that there are more than $n/2$ rectangles whose projection onto l intersects the projection of any of the remaining rectangles. We call such rectangles *l-suitable*. Similarly, we find more than $n/2$ m-suitable rectangles. By the pigeonhole principle, there is a rectangle which is both l-suitable and m-suitable. It is easy to see that it intersects all other rectangles.

7.1.10. Consider the square that is at least as high (in the plane) as any other square. No more than $2(k-1)$ squares intersect this square (each of them contains one of the two lower vertices). Next, as in problem 7.1.3, we apply induction on the number of squares.

7.1.11. Put the nails at the vertices of some square lattice and project onto l and m. Then use the ideas of the solution of problem 7.1.4.

7.1.12. The projections of the centers of the squares onto l can be covered by a segment AB of length 2. Therefore, the unit segment Δ_l, the midpoint of which coincides with the midpoint of AB, intersects the projections of all squares onto l. Similarly, there is a unit segment Δ_m, intersecting the projections of all squares onto m. A square whose projections onto l and m intersect Δ_l and Δ_m, respectively, intersects all the others.

For the following three problems, *the pigeonhole principle for areas* is useful; see problem 7.5.3 in Section 5 of this chapter.

7.1.13. "Inflate" each square A by a factor of 2; that is, consider the square A' obtained by homothety with center at the center of A and coefficient 2. Note that squares A and B intersect if and only if A' contains the center of the square B. Consider the 19×19 square obtained by removing a $1/2$-wide strip along the perimeter of the table. The area covered by inflated squares does not exceed $4 \cdot 90 = 360 < 361 = 19^2$. Thus there must be a point in the 19×19 square that is not covered by any inflated square. This point is the center of a unit square which will not intersect any other square.

7.1.14. Apply the idea used in the solution to problem 7.1.13. You need to modify the definition of "inflation" for a unit square to include all points located at a distance of no more than $1/2$ from some point of the square (this is a square 2×2 with rounded corners). It is easy to verify that the inequality $95 s_0 < 19^2$ is true, where $s_0 = \frac{\pi}{4} + 3$ is the area of the "inflated" unit square.

7.1.15. It is easy to prove that the sum of the lengths of the projections of any rectangle onto l and m is not less than the half-perimeter of the rectangle. It suffices to show that some point of l or m is covered by projections of at least six rectangles. If this were not so, then the sum of the lengths of all the projections would not exceed $5 \cdot (20 + 20) = 200$, a contradiction.

7.1.17. *Answer*: not possible.

Suggestion. Suppose there exists the cycle $P_1 P_2 \ldots P_{12}$. Then a pair of parallelepipeds (P_1, P_2) projects onto one of the three axes as a pair of disjoint segments. The same is true for pairs (P_4, P_5), (P_7, P_8), (P_{10}, P_{11}). In projections onto one of the three axes, we have two pairs of disjoint segments $\Delta_1 \cap \Delta_2 = \emptyset$, $\Delta_3 \cap \Delta_4 = \emptyset$ for which the intersections $\Delta_i \cap \Delta_j$ $(i < j,\ (i,j) \neq (1,2),(3,4))$ are not empty. It remains to solve a one-dimensional problem: to make sure that this configuration of four segments on a straight line is impossible.

7.1.18. Consider the projection onto an arbitrary line and use problem 7.1.1.

7.1.19. *Answer*: it is impossible.

Suggestion. Project the edges a, b, c of the inner parallelepiped onto edges x, y, z of the external one. Estimate the total sum $s = a_x + a_y + \cdots + c_z$ of these projections. On the one hand, $a_x + a_y + a_z \geq a$, $b_x + b_y + b_z \geq b$, $c_x + c_y + c_z \geq c$, whence $s \geq a + b + c$. On the other hand, it is easy to show that $a_x + b_x + c_x \leq x$, $a_y + b_y + c_y \leq y$, $a_z + b_z + c_z \leq z$. Therefore, $s \leq x + y + z$.

7.1.20. *Answer*: it is possible.

2. Helly's theorem (2)
By A. V. Akopyan

Before studying this section, it is helpful to solve problems 7.1.1 and 7.1.2 in the previous section.

A subset A of the plane is called *convex* if for any two of its points it contains the line segment connecting them. All figures discussed below are assumed to be bounded.

7.2.1. Hahn-Banach theorem. (a) Suppose we are given two disjoint convex polygons on a plane. Prove that there exists a line that "separates" them: the line does not intersect either polygon and is such that the polygons lie on different sides of the line.

(b)* Generalize this theorem to the case of n-dimensional space.

7.2.2. Helly's theorem. (a) Let every three polygons from a set of convex polygons have at least one common point. Prove that then all the polygons in this set have a common point.

(b)* Generalize this theorem to the case of n-dimensional space.

7.2.3.° Which of these theorems cease to be true if we allow nonconvex polygons? Select all valid options:
(1) Hahn-Banach theorem;   (2) Helly's theorem.

7.2.4. (a) Suppose we are given a system of arcs on a circle, each arc shorter than the length of a semicircle. It is known that every three arcs in this set have at least one common point. Prove that all the arcs have at least one common point.

(b) What is the maximum length of arcs for which the pairwise condition of the intersection of the arcs is sufficient for the existence of a common point of all arcs in the system?

7.2.5. (a) Prove that if every three points of some subset of the plane can be covered by a circle of radius R, then all points of this subset can be covered with a circle of the same radius.

(b)* Prove that if every three lines from some set of lines can be intersected by a circle of radius r, then all the lines from this set can be intersected by a circle of radius r.

7.2.6. Is it possible to generalize Helly's theorem to *pairs* of points? In other words, the question is about the existence of a number n with the following property. Suppose we are given a family of convex polygons on a plane such that for every n polygons from this family there exist two points such that a convex polygon in the subfamily contains at least one of them. Then do there exist two points such that every polygon in the family contains at least one of them?

7.2.7. 100 sets A_1, A_2, ..., A_{100} are selected on the line, each of which is a union of 100 pairwise disjoint segments. Prove that the intersection of the sets $A_1, A_2, \ldots, A_{100}$ is a union of not more than 9901 pairwise disjoint segments. (In this problem, a point is considered to be a segment.)

7.2.8. A line contains $2k-1$ white and $2k-1$ black segments. It is known that any white segment intersects at least with k black segments and any black segment intersects at least with k white segments. Prove that there is a black segment intersecting with all white ones and a white segment intersecting all black ones.

7.2.9. On a rectangular table lie congruent cardboard squares of k different colors with sides parallel to the sides of the table. The squares are arranged so that in any set of k of them, there are two of these that can be nailed to the table with a single nail. Prove that all the squares of some color can be nailed to the table using $2k-2$ nails.

7.2.10. A finite set of points X and an equilateral triangle T in the plane are given. It is known that any subset X' of the set X consisting of no more than 9 points can be covered by two parallel translations of the triangle T. Prove that all the points in X can be covered by two parallel translations of the triangle T.

Suggestions, solutions, and answers

7.2.5. (a) (A. Bezmenova) Draw circles of radius R with center at each point. Since any three points can be covered by a circle of radius R, the radius of the circumscribed circle of any triangle with vertices at these points is less than R. Therefore, circles with centers at the vertices of one triangle intersect, that is, any three circles intersect, and, therefore, by Helly's theorem, all the circles intersect. A circle of radius R with center at a point common to all circles will cover all the points.

    7.2.8. (A. Rubashevsky) We give a solution for the special case when same-color segments do not intersect. To begin with, we prove that *for each pair of black segments, there is a white one that intersects both of them.* Consider two arbitrary black segments. By hypothesis, each black segment intersects at least k white segments. The first black segment intersects at least k white segments, and the second segment intersects at least k white. But since there are only $2k - 1$ white segments, by the pigeonhole principle there is at least one white segment intersecting both of them.

    Now take the following two black segments: one whose left end is rightmost and the other whose right end is leftmost. Denote these segments by (a_1, a_2) and (b_1, b_2), respectively. Note that these black segments are extremal in the sense that for every other black segment (x_1, x_2) we have $x_1 \leq b_1$, $x_2 \geq a_2$. By the statement proved above, for these two selected segments there is at least one white segment (z_1, z_2) that intersects both of them, so that $z_1 \leq a_2$, $z_2 \geq b_1$. Therefore, for each black segment (x_1, x_2) we have $x_1 \leq z_2$, $x_2 \geq z_1$. This means that the white segment (z_1, z_2) intersects all black segments. Similarly we can show that there is a black segment that intersects all white segments.

3. Lattice polygons (2)
By V. V. Prasolov and M. B. Skopenkov

This series of problems is devoted to two beautiful theorems about lattice polygons (also known as "grid paper polygons"): Pick's formula (see 7.3.6) for the area of such polygons and the more complicated dual polygon theorem (twelve-point theorem). We present a proof of the first result following [**Vas74**] and a proof of the second result following [**CRM**].

    Lattice polygons arise naturally in determining the number of solutions of systems of algebraic equations. This is explored in the interesting articles [**Kush1**] and [**Kush2**]. The concept of duality (relevant to the twelve-point theorem) and its connection with the solution of equations can be found in the article [**Tab**] and in the more advanced article [**PR-V**].

For Pick's formula, you should be familiar with areas and the extremal principle (Section 2 of Chapter 2 in this book and Section 5.3 in [**ZS**]). For the twelve-point theorem, familiarity with semi-invariants (Section 4 in Chapter 4 in this book) and polarity (Section 5.3 in [**ZS**]) is also helpful.

In this section, we assume that all polygons are lattice polygons (the vertices are lattice points of a unit grid).

3.A. Area of a polygon on grid paper (2)

Our immediate goal is to find a simple way to calculate the area of a lattice polygon. We start with special cases.

7.3.1.° Find the areas of the polygons in Fig. 1.

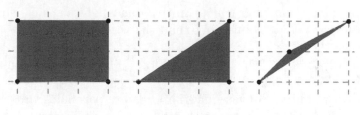

FIGURE 1

We call a lattice triangle *primitive* if it contains no other lattice points besides its vertices (neither inside nor on its sides). (See the rightmost triangle in Fig. 1.) A *leap* (Fig. 2) is the transformation in which the vertex A of the triangle ABC is replaced by a point symmetric to A with respect to B.

7.3.2.° What is the longest side of a primitive triangle?
  (1) 1;   (2) $\sqrt{2}$;   (3) 2;   (4) $\sqrt{13}$;   (5) arbitrarily large.

7.3.3. Prove that
  (a) the area of a triangle does not change after a leap;
  (b) a primitive triangle remains primitive after a leap;
  (c) a primitive triangle is either obtuse or right (and the latter case is possible only for a triangle with sides 1, 1, $\sqrt{2}$, which we will call *minimal*; see Fig. 2 on the left);
  (d) for any primitive nonminimal triangle, there is a leap that results in a triangle whose largest side is strictly smaller than the longest side of the original triangle;
  (e) any primitive triangle can be translated into a minimal one by a finite number of leaps;
  (f) the area of a primitive triangle is 1/2.

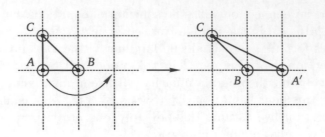

FIGURE 2

The last statement above is the key to solving many other beautiful problems about polygons.

7.3.4.° How many noncongruent primitive quadrilaterals (i.e., quadrilaterals containing no lattice point other than the vertices, neither inside nor on the sides) are there?
    (1) 6;   (2) 7;   (3) 8;   (4) 9;   (5) 12;   (6) infinitely many.

7.3.5. (a) All primitive convex quadrilaterals are parallelograms.
    (b) Is there a convex pentagon that does not contain any interior lattice points?
    (c) A convex polygon has exactly one interior lattice point. What is the largest number of vertices it can have?
    (d) If there are no lattice points on the sides of a triangle (except for the vertices), but there is exactly one interior lattice point, then this point is the intersection point of the medians of the triangle.
    (e) Given a primitive parallelogram, we are allowed to take any side of it and translate it by a vector "equal" to this side. Prove that using such operations one can transform a primitive parallelogram into a unit square.

We will now find the area S of a polygon with i interior lattice points and b lattice points on the boundary (which includes the vertices).

7.3.6. (a) Consider a triangulation of a polygon with vertices at these $b + i$ lattice points. How many triangles will it have?
    (b) **Pick's formula:** $S = i + \frac{b}{2} - 1$.

The following problems are devoted to applications of Pick's formula.

7.3.7. (a) There are no equilateral triangles whose vertices are lattice points.
    (b)* What regular polygons are lattice polygons?
    (c)* For which integers n is the number $\cos n°$ rational?

7.3.8. A chess king traveled on an 8×8 chessboard visiting each square exactly once and returning to the original square. The broken line that connects the centers of the squares that the king visited has no self-intersections.

  (a) What is the area that this broken line encloses?

  (b) What is the greatest length it can have?

7.3.9.* Draw a closed nonself-intersecting broken line along the grid lines which passes through all the interior lattice points of a $p \times q$ lattice rectangle.

  (a) For what p and q is this possible?

  (b) How long is this broken line?

  (c) What is the area of the figure it bounds?

7.3.10.* (a) Let kM be the polygon obtained from M by homothety with coefficient k and center at the origin. Prove the formula $2S(M) = n(2M) - 2n(M) + 1$, where $S(M)$ is the area of the polygon M and $n(M)$ is the number of lattice points inside and on the boundary of the polygon M.

  (b) Guess and prove a similar formula for the volume of a polytope in space.

7.3.11 (Challenge).*** Come up with an estimate for the maximum number of vertices of a convex polygon in terms of the number of interior lattice points

3.B. Dual lattice polygons (3*)

Consider a convex lattice polygon $M = A_1 A_2 \ldots A_n$ (Fig. 3 on the left) that contains exactly one interior lattice point O. Draw vectors $\overrightarrow{A_1 A_2}, \overrightarrow{A_2 A_3}, \ldots,$ $\overrightarrow{A_n A_1}$ from O, and on each of the resulting segments, select the lattice point closest to O. Connecting these points consecutively, we obtain a new polygon M^*, which is called *dual* to the original polygon (Fig. 3 on the right).

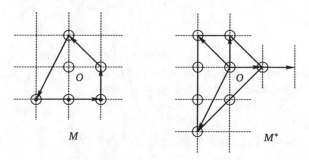

$$M \qquad\qquad\qquad\qquad M^*$$

FIGURE 3

7.3.12.* The dual polygon is identical to the polygon obtained by taking the polygon polar to M in the circle of radius 1 centered at O and rotating it by $90°$ counterclockwise around O.

Our goal is to obtain an elementary proof of the following very recently discovered result (see [**PR-V**]).

Theorem (twelve-point theorem). *Suppose a convex polygon M has exactly one interior lattice point O and b lattice points on its boundary, and suppose its dual M^* has b^* lattice points on its boundary. Then*

$$b + b^* = 12.$$

7.3.13. Draw the polygons dual to those shown in Fig. 4. How many nodes are located inside M^* in each case? What is M^{**} equal to? How are the areas M and M^* connected? Is the equality $M = M^*$ possible? Can the dual to a convex polygon be nonconvex? Formulate your observations and assumptions, and try to prove them.

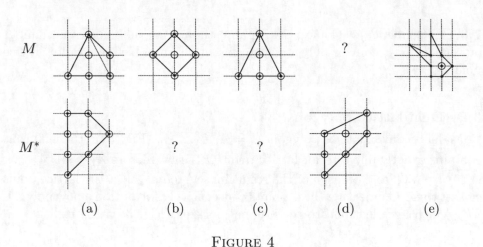

$$\text{Figure } 4$$

The next problem applies the twelve-point theorem. First try to solve it using this theorem without proof, and then try to solve it without using this theorem. See also problem 7.3.5(c) above.

7.3.14. Inside
    (a) a triangle; (b) a convex quadrilateral
there is exactly one interior lattice point. What is the largest number of boundary lattice points?

7.3.15. Prove the twelve-point theorem for:
    (a) a parallelogram with $b = 4$;   (b) a triangle with $b = 3$.

Definition. *Deleting a triangle* from M is the operation of removing from M a primitive triangle with *two sides contained in the sides of M*. For example, deleting a triangle from the polygon in Fig. 4(a) results in the polygon in Fig. 4(c). The inverse operation is called *adding a triangle*.

7.3.16. Is it true that any two polygons
 (a) in Fig. 4;
 (b) with no interior lattice points;
 (c) that are convex, with exactly one interior lattice point
are obtained from each other by a series of deletions and additions of triangles?

7.3.17. (a) When a triangle is deleted from M, a triangle is added to M^*.
 (b) Prove the twelve-point theorem.

7.3.18.* Prove that if M has exactly one interior lattice point, then M^* also has exactly one interior lattice point.

7.3.19.* The polygon M^{**} is obtained from M by central symmetry.

An *affine lattice transformation* is a mapping of the plane given by the formula
$$\begin{cases} x' = ax + by + e, \\ y' = cx + dy + f \end{cases}$$
that maps the set of lattice points to itself (this means that the numbers a, b, c, d, e, f are integers and $ad - bc = \pm 1$). Two polygons are called *equivalent* if there exists an affine lattice transformation mapping one polygon into another.

7.3.20.* Prove that among
 (a) parallelograms with area equal to 9;
 (b) polygons with exactly one interior lattice point
there are only a finite number of pairwise nonequivalent polygons.

Suggestions, solutions, and answers

7.3.1. *Answer*: 6, 3, 1/2 (from left to right).
 7.3.2. *Answer*: (5).
 7.3.3. (a) The triangles ABC and $A'BC$ have a common base BC, and since $AB = BA'$, the corresponding altitudes are equal, so the areas of these triangles are equal.
 (b) Let D be the midpoint of BC. Note that under a central symmetry with center D the lattice is transformed into itself. Therefore, if the triangle ABC is primitive, its image under this central symmetry $A''BC$ is also primitive. Similarly, the triangle $A''BA'$ is primitive because it is obtained

by central symmetry from the triangle $A''BC$. Therefore, there are no lattice points inside the parallelogram $A''A'BC$; in particular, triangle $A'BC$ is primitive.

(c) Consider a minimal rectangle with sides parallel to the grid lines containing our triangle. Then on each side of it lies some vertex of the triangle. This is possible only if at least one of the vertices of the triangle coincides with a vertex of the rectangle. If there are three such coincident vertices, then it is easy to see that our triangle is minimal. If there are two of them and they are opposite (in the rectangle), then, obviously, the angle at the third vertex of the triangle is obtuse. The other cases for a primitive triangle are impossible. Indeed, suppose that a triangle and a rectangle have a common vertex A, but either there are no more coinciding vertices or a common vertex is one of the vertices of the rectangle adjacent to A. Let D be the vertex of the rectangle opposite to A, and let M be the midpoint of the side of the triangle opposite to A. Then it is easy to see that the point symmetric to D with respect to M is a lattice point lying inside the original triangle or on its side.

(d) It suffices to leap over the vertex B with an obtuse angle. Since AC is the largest side, we have $AC > BC$ and $AC > AB = BA'$. Since the angle ABC is obtuse, it is larger than the angle $A'BC$, which means that $AC > A'C$ (because the other two sides of the triangles ABC and $A'BC$ are equal).

(e) The leaping process reduces the length of the largest side of primitive triangles. This process cannot continue infinitely, since the square of the length of this side takes only integer values (by the Pythagorean theorem). By (d), this process will stop at the minimum triangle.

(f) The statement follows from parts (a) and (e).

7.3.4. *Answer*: 6.

7.3.5. (a) Let $ABCD$ be our quadrilateral, $O = AC \cap BD$. Since $ABCD$ contains no interior or boundary lattice points, triangles ABC and ACD are primitive. By 7.3.3(f), their areas are equal to $1/2$, and, therefore, $BO = OD$. Similarly, $AO = OC$. Therefore $ABCD$ is a parallelogram.

(b) *Answer*: no.

Suggestion. Without loss of generality we can assume that there are no lattice points on the sides of the pentagon $ABCDE$ (otherwise, we consider a smaller pentagon). Then, by (a), $ABCD$ is a parallelogram, and $BCDE$ is also a parallelogram. It follows that $A = E$, which is a contradiction. Hence, the required pentagon does not exist.

(c) *Answer*: 6.

Suggestion. An example is shown in Fig. 5(e) of this chapter. Let us prove maximality (this also follows from the twelve-point theorem). Draw a line through the interior lattice point O that does not pass through its vertices. It separates the plane into two regions. If the number of vertices is 7 or more, then by the pigeonhole principle in one of the regions there are at least 4 vertices A, B, C, D. Then the pentagon $OABCD$ contains no interior lattice points, contradicting (b). This proves maximality.

From this, in particular, we see that the number of vertices for the polygon M in the twelve-point theorem can be $n = 3, 4, 5, 6$.

(d) Let the triangle ABC contain a single interior lattice point O. Then triangles AOB, BOC, and COA are primitive, with area $1/2$ by 7.3.3(f). In particular, this means that the line AO is equidistant from the vertices B and C. Thus AO is a median, as are BO and CO.

(e) The statement follows from 7.3.3(e).

7.3.6. (a) *Answer*: $2i + b - 2$.

Suggestion. Calculate the sum of the angles of the triangles in the triangulation in two ways. On one hand, it is equal to $n\pi$, where n is the number of triangles. On the other hand, each interior lattice point contributes 2π to this sum. The sum of the angles at the boundary lattice points contributes an amount equal to the sum of the angles of a b-gon (for which some angles could be π), which is $(b-2)\pi$. Thus $n\pi = 2i\pi + (b-2)\pi$.

(b) Pick's formula follows from part (a), problem 7.3.3(f), and the well-known proposition that every polygon can be dissected into triangles.

This proof is discussed in more detail in [**Vas74**]. Here is a sketch of a simpler proof (in comparison with the solution to problem 7.3.3(f); problem 7.3.3(f) itself is a consequence of Pick's formula). Let $\varphi(P, M)$ denote the angle at which the polygon M is "visible" from the lattice point P; i.e.,

$$\varphi(P, M) = \begin{cases} \angle P, & \text{if } P \text{ is a vertex,} \\ \pi, & \text{if } P \text{ lies on a side,} \\ 2\pi, & \text{if } P \text{ lies inside } M. \end{cases}$$

Denote

$$\varphi(M) = \sum_{P \in M} \varphi(P, M),$$

where the sum is taken over all lattice points inside and on the boundary of M. It suffices to prove the following statements:

(1) $\varphi(M) = (2i + b - 2)\pi$;

(2) $\varphi(M \cup N) = \varphi(M) + \varphi(N)$ if the polygons M and N do not have common interior points;

(3) $\varphi(M) = 2\pi S(M)$ for the following polygons M:

(a) M is a rectangle with sides parallel to the grid lines;

(b) M is a right triangle with legs parallel to the grid lines;

(c) M is an arbitrary triangle;

(d) M is an arbitrary polygon.

7.3.7. (a), (b) Let the ratio of the area of the polygon to the square of one of its sides be irrational (as is the case, for example, for an equilateral triangle). Then this polygon cannot be a lattice polygon.

Indeed, if the vertices of the polygon lie on lattice points, then the square of the side is an integer (by the Pythagorean theorem) and the area is a half-integer (by Pick's formula); therefore their ratio is rational.

7.3.8. (a) *Answer*: 31.

Suggestion. Since all squares are visited, Pick's formula implies that the area enclosed by the broken line is equal to $64/2 - 1 = 31$ (letting the lattice points be the centers of the chessboard squares).

(b) *Answer*: $28 + 36\sqrt{2}$.

First suggestion. It is easy construct a path in which 36 of 64 moves are $\sqrt{2}$ long (traveling diagonally). We will show that there can be no more than 36 such moves. For each path segment of length $\sqrt{2}$, consider the unit square that has this segment as a diagonal. One half of this square lies outside the polygonal region enclosed by the king's path. But the total area occupied by these halves does not exceed $49 - 31 = 18$, since they do not go beyond the boundaries of a 7×7 square. Hence, the number of diagonal moves does not exceed 36.

Second suggestion (A. Tolmachev). Call a chessboard square a "boundary" square if it is adjacent to the edge of the board. Thus the chessboard contains 28 boundary squares. Start at a boundary square and consider the first time when we again return to the boundary (possibly after several moves). At this moment, we could only be in the square adjacent to the original one, since the path is closed and does not have self-intersections. Since neighboring boundary squares on a chessboard have different colors, there will be one horizontal or vertical move when moving from one of the neighboring cells to another (because if you walk only diagonally, the color will not change). And since there are a total of 28 boundary cells, there will be at least 28 horizontal or vertical moves.

7.3.9. *Answer*: (a) for $p, q \geq 3$ and $(p-1)(q-1)$ even;

(b) $(p-1)(q-1)$;

(c) $\frac{(p-1)(q-1)}{2} - 1$.

Suggestion. Use parity considerations and Pick's formula.

7.3.10. Part (a) immediately follows from Pick's formula (see problem 7.3.6(b)) and can also be proved similarly to the proof of Pick's formula (see the remark after the solution to problem 7.3.6).

For (b), several different formulas can be proposed; for example,

$$6V(M) = n(3M) - 3n(2M) + 3n(M) - 1]$$

or

$$6V(M) = n(2M) - 2n(M) - b(M) + 3.$$

These formulas are proved in the interesting article [**Kush1**], which also explains how all such formulas can be obtained. The article [**Kush2**] by the same author is also recommended.

7.3.11. Let $n(i)$ denote the maximum number of vertices of a convex polygon that has i interior lattice points. Here are the first few values of

$n(i)$:

i	0	1	2	3	4	5	...
$n(i)$	4	6	6	6	8	7	...

The formulas $n(0) = 4$ and $n(1) = 6$ are actually proved in problem 7.3.5(b), (c). To prove the equality $n(2) = 6$, it suffices to draw a straight line through two interior lattice points and then use ideas from the solution to problem 7.3.5(c).

For bigger i we can obtain the following estimate for the number $n(i)$:

$$n(i) \leq 2i \quad \text{for} \quad i \geq 3.$$

For a proof, we consider the convex hull of the i interior lattice points. This is either a line segment or a polygon with at most i vertices. In the first case, using the ideas of 7.3.5(c), it is easy to obtain the estimate $n(i) \leq 6$. In the second case, take an arbitrary point O (not necessarily a lattice point) inside the convex hull and draw rays from it through all the vertices of the convex hull. By choosing the point O, we can ensure that the vertices of the original polygon do not fall on our rays. If the number of vertices is $n(i) > 2i$, then, by the pigeonhole principle, one of the regions partitioned by the rays must contain at least three vertices. Together with two vertices of the convex hull belonging to this region, they form a convex pentagon without interior lattice points, contradicting 7.3.5(b).

The following estimate is also known (but the proof is complicated):

$$n(i) \leq Ci^{1/3},$$

for some constant C.

Note the interesting fact that $n(5) < n(4)$.

7.3.12. Use the following statement, which follows from 7.3.3(f): O is the only interior lattice point of M if and only if, for any two neighboring lattice points A and B on the boundary of the polygon M, the area of the triangle AOB is $\frac{1}{2}$.

7.3.13. See Fig. 5(a)–(d). The dual to the polygon of Fig. 4(e) is not defined since this polygon is not convex. It is also easy to verify that for a convex polygon, the dual polygon is also convex. Fig. 5(e) shows an example where $M = M^*$. Here are some observations:

(1) exactly one lattice point is located inside M^* (see 7.3.18);

(2) M^{**} is obtained from M by central symmetry (see 7.3.19);

(3) $S(M) + S(M^*) = 6$, which is equivalent to the twelve-point theorem, since by Pick's formula, $S(M) = i + \frac{b}{2} - 1 = \frac{b}{2}$.

7.3.14. (a)*Answer*: 9.

Suggestion. An example is shown in Fig. 5(f). The idea for the example is suggested by the twelve-point theorem. Since 12 lattice points in total are located on the boundaries of the triangles T and T^*, to maximize the number of lattice points on the boundary of T, one must minimize their number on the boundary of T^*, and vice versa. Therefore, we need to take

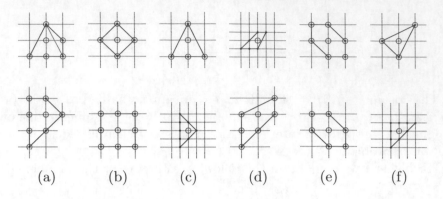

(a)   (b)   (c)   (d)   (e)   (f)

FIGURE 5

a triangle with 3 lattice points on the boundary and then the dual triangle
will be the desired example.

The maximality of the number 9 follows, of course, from the twelve-
point theorem. We give an independent proof (Fig. 6). If on each side of a
triangle there are no more than 2 lattice points (not counting the vertices),
then there is nothing to prove. Consider the case when one of the sides
of the triangle, say, AB, contains at least 3 lattice points (such a situation
is indeed possible; see Fig. 5(c)). It is enough to prove that on each of
the other sides there is at most one lattice point. Suppose that this is not
so and, for example, AC contains at least 2 lattice points. Among them,
choose the lattice point D closest to A. Draw $DE \parallel AB$, where $E \in BC$.
On side AB, choose the lattice point F closest to A. Consider the points
G and H on the segment DE such that $DG = AF$ and $GH = AF$. They
are obviously lattice points. Since the lattice points divide AB and AC into
equal segments, we have $CD : CA \geq \frac{2}{3}$ and $AF : AB \leq \frac{1}{4}$. Using similar
triangles, we have $DE : AB \geq \frac{2}{3}$. It follows that $DH < DE$; i.e., G and H
lie inside the triangle, a contradiction, proving that 9 is the maximum.

(b) *Answer*: 8.

Suggestion. An example is shown in Fig. 5(b).

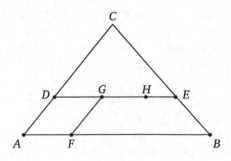

FIGURE 6

7.3.15. (a) *Solution* (see [**CRM**, proof of statement (1)]). Let $M =$
$ABCD$ be a parallelogram. Then $O = AC \cap BD$, since the point symmetric

to the point O with respect to $AC \cap BD$ is a lattice point and lies inside $ABCD$ and therefore equals O. It is easy to see that M^* is a parallelogram whose sides are obtained from the diagonals AC and BD by parallel translations by the vectors $\pm \overrightarrow{OB}$ and $\pm \overrightarrow{OA}$, respectively. Since there is a single lattice point O on these diagonals, on each side of the parallelogram M^* lies one lattice point that is not a vertex. Therefore, $b + b^* = 4 + 8 = 12$.

(b) By 7.3.5(d), the lattice point O is the intersection point of the medians of triangle ABC. It is well known that

$$
\begin{cases}
\overrightarrow{AB} - \overrightarrow{CA} = 3\overrightarrow{AO}, \\
\overrightarrow{BC} - \overrightarrow{AB} = 3\overrightarrow{BO}, \\
\overrightarrow{CA} - \overrightarrow{BC} = 3\overrightarrow{CO}.
\end{cases}
$$

However, the vectors of the left sides are the vectors of the sides of the dual triangle. Therefore, the dual triangle is obtained from the triangle constructed by the vectors $\overrightarrow{AO}$, $\overrightarrow{BO}$, and $\overrightarrow{CO}$ by a homothety with coefficient 3. Since segments AO, BO, and CO do not contain lattice points (besides endpoints), each side of the dual triangle contains exactly 2 lattice points (except for the vertices). Therefore, $b + b^* - 3 + 9 = 12$.

7.3.16. The answer to all three questions is yes. Part (a) follows from (c).

(b) It is sufficient to prove that starting with a given polygon without interior lattice points and using a sequence of additions and deletions of *triangles* one gets a minimal primitive triangle (see 7.3.3(c)). Consider the triangulation of a polygon whose vertices are all of its boundary lattice points. All triangles in the triangulation will be primitive. Among them, *by problem* 3.1.4, there is an "extreme" triangle, i.e., a triangle sharing two sides with the original polygon. Delete it. Continue this process until a single primitive triangle remains. It remains to use problem 7.3.3(e) since the leap of a primitive triangle is the result of adding a triangle and subsequently deleting the original triangle.

(c) First we prove a lemma:

Lemma (see [**CRM**, statement (3)]). *Starting with any polygon M, by a series of deletions/additions of triangles, we can obtain a parallelogram with $b = 4$ and $i = 1$.*

Proof. Consider all lattice points on the sides of the polygon M as vertices (possibly with angle of 180°). This will not alter the definition of M^* or the processes of deletions/additions of triangles.

Assume first that M has a diagonal not passing through O. Cut M along this diagonal and consider the part that does not contain O. This part necessarily contains a primitive triangle of the form $A_{k-1} A_k A_{k+1}$. Therefore, by removing this triangle, you can reduce the value of b. Continue as long as possible. Obviously, there are only three cases where the required diagonal is not found:

(1) $b = 4$, $M = ABCD$ is a quadrilateral, $O = AC \cap BD$. Since the segments OA, OB, OC, and OD do not contain lattice points, we have $OA = OC$ and $OB = OD$; that is, $ABCD$ is the desired parallelogram.

(2) $b = 4$, $M = ABCD$ is a triangle, one of the angles, say, BCD, is straight. In this case, let D' be the point symmetric to D with respect to O, and let E be the midpoint of $D'B$. The desired series of operations has the form

$$ABCD \to AEBCD \to AD'EBCD \to AD'ECD \to AD'CD \quad \text{(Fig. 7)}.$$

(3) $b = 3$, $M = ABC$. In this case, denote by A' and C' points symmetric with respect to O to the vertices A and C, respectively. Then the desired series has the form

$$ABC \to AC'BC \to AC'BA'C \to AC'A'C \quad \text{(Fig. 7)},$$

proving the lemma.

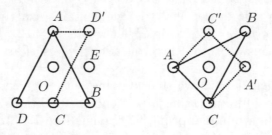

FIGURE 7

To solve the problem it suffices to prove that starting with any parallelogram with $b = 4$, $i = 1$, a series of deletions/additions of triangles will result in the parallelogram shown in Fig. 4(b). Let $ABCD$ be our parallelogram. Triangle AOB is primitive. If it is minimal, then our parallelogram is congruent to the parallelogram in Fig. 4(b). Otherwise, consider the sequence of leaps of triangle AOB which transforms it into a minimal triangle (see 7.3.3(f)), and at each step, we will "double" triangle AOB to form a parallelogram. It is easy to see that any two of the successive parallelograms thus formed can be obtained from each other by a series of deletions/additions of triangles. Our assertion is proved.

7.3.17. (a) *Solution* (see [**CRM**, proof of statement (2)]). Again, consider all lattice points on the sides of M to be vertices (possibly with angle 180°). Without loss of generality, assume that the primitive triangle $A_1 A_2 A_3$ is deleted from M. Let us prove that this leads to the addition of a primitive triangle $A_{12} A_{13} A_{23}$ to M^*, where the point denoted by A_{kl} means $\overrightarrow{OA_{kl}} = \overrightarrow{A_k A_l}$ (Fig. 8). In particular, if $l = k+1$, then A_{kl} is a vertex of the polygon M^*.

Delete triangle $A_1A_2A_3$. Then vertices A_{12} and A_{23} disappear from M^*, but a new vertex A_{13} will be added. It still needs to be joined by segments to A_{n1} and A_{34}. We show that points A_{12} and A_{23} lie on these segments. Indeed, since O is the only lattice point inside M, then triangles A_1OA_3, A_2OA_3, and A_4OA_3 are primitive. By Pick's formula, their areas are equal to $1/2$. Since they have a common base OA_3, the projections of $\overrightarrow{A_1A_3}$, $\overrightarrow{A_1A_3}$, and $\overrightarrow{A_1A_3}$ onto the perpendicular to OA_3 are equal. It follows that points A_{13}, A_{23}, and A_{34} lie on one line, and since M is convex, A_{23} lies between the other two. Similarly, we can show that A_{12} lies on the segment $A_{n1}A_{13}$. Therefore, the transformation of M^* is reduced to adding triangle $A_{12}A_{13}A_{23}$ to it.

Note that the triangle $OA_{12}A_{13}$ is obtained from the primitive triangle $A_1A_2A_3$ by parallel translation, and $OA_{23}A_{13}$ is obtained by central symmetry. Therefore the triangle $A_{12}A_{13}A_{23}$ is primitive, as required.

(b) From part (a) we see that when deleting/adding a triangle, the value $b + b^*$ is preserved. According to the lemma used in the solution to problem 7.3.16(c), this polygon can be converted into a parallelogram with $b = 4$ by a sequence of deleting/adding triangles. For such a parallelogram, we have $b + b^* = 12$ (see 7.3.15(a)). The twelve-point theorem is proved.

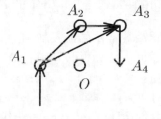 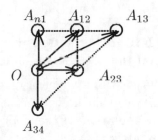

FIGURE 8

7.3.18. Suppose that inside M^* there is an interior lattice point Q different from O. Then it is located inside some triangle AOB where A and B are a pair of neighboring vertices of the polygon M. By Pick's formula, the area of this triangle is not less than $3/2$. Moreover, it can equal $3/2$ only if Q lies on the median OK, by problem 7.3.4(4). By the definition of M^*, the boundary of M contains three lattice points D, E, and F such that $\overrightarrow{DE} = \overrightarrow{OA}$ and $\overrightarrow{EF} = \overrightarrow{OB}$. The area of triangle DEF is equal to the area of the triangle ABC; i.e., it is not less than $3/2$. If it is greater than $3/2$, then Pick's formula implies that there are at least two lattice points either inside DEF or on the interior of side DF. If the area is equal to $3/2$, then $DF \parallel OK \parallel OQ$. But $DF = 2OK > 2OQ$, which again implies that there are at least two lattice points in the interior of the side DF.

Since M is a convex polygon, there are two possibilities: either DF lies inside M or DF is part of the boundary of the polygon M. In the first case, M must contain at least two interior lattice points. In the second case, $M = DEF$ also cannot contain exactly one interior lattice point. Indeed, M is divided into three or more triangles with vertex D obtained from each other by leaps; using problem 7.3.3(b), either all these triangles are primitive or they are all not primitive. We get a contradiction. Therefore, such a lattice point Q cannot be found.

7.3.19. Use 7.3.12.

7.3.20. (b) The proof is based on the same idea as the classification of primitive polygons (see [**PR-V**]). The classification of convex polygons with exactly one interior lattice point is discussed in [**Kh**] (but without proof). It turns out that there are only 16 such polygons up to an affine transformation of the lattice.

4. Pigeonhole principle on a line (3)
By A. Ya. Kanel-Belov

7.4.1. Holes with width 0.01 are located on the number line, centered at each integer point. A hare jumps along the line, starting at zero, with jump length of $\sqrt{2}$. Prove that sooner or later the hare will fall into a hole.

7.4.2. Kronecker's density lemma. (a) At each point of the integer lattice on the plane sits a hare (a hare is a circle of radius 0.01). A hunter standing at the point $\left(\frac{1}{2}, \frac{1}{2}\right)$ shoots in a direction that has an irrational slope. Prove that he will hit at least one hare.

Reformulation: if the slope is irrational, then the line comes arbitrarily close to lattice points.

(b) If the slope is rational, then, for a sufficiently small size of the hare, the hunter can be placed so that he will not hit any hares.

(c) Flowers (i.e., circles of radius 0.01 with centers at the lattice points) are planted on a (plane) field. A horse gallops the field along a straight line in one direction leaping repeatedly by the vector $(\sqrt{2}, \sqrt{3})$. Prove that he must knock down at least one flower.

7.4.3. (a) Prove that the number $\log_{10} 2$ is irrational.

(b) Prove that the number 2^n may start with any combination of digits (for an appropriate n).

7.4.4. Find the probability that 2^n starts with the digit 1. More rigorously: let a_n denote the number of exponents k between 1 and n for which 2^k starts with 1. Prove that limit $\lim_{n \to \infty} \frac{a_n}{n}$ exists, and find it. (The limit definition used here is given in [**Sko**, §7.3, "Concrete limit theory"].)

7.4.5. Prove that there are infinitely many integers m and n such that $\left|\sqrt{2} - \frac{m}{n}\right|$ is less than

(a) $\frac{1}{2n}$;   (b) $\frac{1}{999n}$;   (c) $\frac{1}{n^2}$;   (d)* $\frac{1}{\sqrt{5}n^2}$.

7.4.6.* (a) For any $\varepsilon > 0$ there exists a countable set of intervals, the sum of whose lengths is less than ε, such that, if α is not in any of the intervals and c is any positive number, there exist $m, n \in \mathbb{Z}$, $n > 0$, such that $\left|\alpha - \frac{m}{n}\right| < \frac{1}{cn^{2+\varepsilon}}$.

(Rigorous reformulation: for any $\varepsilon > 0$, the set of real numbers that are not $(2 + \varepsilon)$-approximated has measure 0.)

(b) **Hurwitz-Borel theorem.** For any irrational number α, there are infinitely many $m/n \in \mathbb{Q}$, such that $\left|\alpha - \frac{m}{n}\right| < \frac{1}{n^2\sqrt{5}}$.

(c) The number $\sqrt{5}$ in the Hurwitz-Borel theorem cannot be increased: for any $c > \sqrt{5}$ there is an irrational number α such that the inequality $\left|\alpha - \frac{m}{n}\right| < \frac{1}{cn^2}$ holds for only a finite number of $m/n \in \mathbb{Q}$,

Suggestions, solutions, and answers

7.4.3. This problem is analyzed in [**Bol78**] and [**Ar98**].

7.4.5. (b) For any positive integers N, k and any irrational number α there are at least k different fractions $m/n \in \mathbb{Q}$ for which $n \leq Nk$ and $\left|\alpha - \frac{m}{n}\right| < \frac{1}{Nn}$. For details, see the suggestion to problem 7.5.14 below.

5. The pigeonhole principle and its application to geometry[1]   (3) *By I. V. Arzhantsev*

The area of a figure

We will call a planar figure A *simple* if it can be cut into a finite number of triangles. Its area $S(A)$ is defined as the sum of the areas of the corresponding triangles.

Recall that a point $(x_0, y_0) \in A$ is called an *interior* point of A if there is a circle with center (x_0, y_0) entirely lying in A.

It is easy to verify that the function "area" on the set of simple figures has the following properties:

• if A has interior points, then $S(A) > 0$;

• if A is the union of simple figures A_1 and A_2 without common interior points, then $S(A) = S(A_1) + S(A_2)$;

• congruent figures have the same area;

• the area of a unit square is 1.

---

[1]Based on [**Yad**].

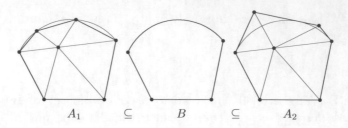

$$A_1 \quad \subseteq \quad B \quad \subseteq \quad A_2$$

FIGURE 9

More generally, a planar set B is called *measurable* if for any $\varepsilon > 0$ there exist simple figures A_1 and A_2 such that $A_1 \subseteq B \subseteq A_2$ and $S(A_2) - S(A_1) < \varepsilon$ (see Fig. 9). For measurable sets, one can also define the concept of area and prove that the area is the only function on the set of measurable sets that has the four properties listed above.

Note that not every plane set is measurable (see, for example, problem 7.5.2). To those who want to learn more about the concept of area and its generalizations we can recommend the book [**Leb**].

7.5.1. Prove that a bounded figure whose boundary consists of a finite number of segments and arcs of circles is measurable.

Recall that a planar set is called *bounded* if it is contained in some circle.

7.5.2. Prove that any measurable set is bounded.

The pigeonhole principle for areas

The following geometric statement resembles the well-known "pigeonhole principle" and is therefore usually called the *geometric pigeonhole principle* or the *pigeonhole principle for areas*.

7.5.3. Pigeonhole principle for areas. Let A be a measurable set, and let $A_1, \ldots, A_m$ be measurable subsets of A. Suppose that

$$S(A) < S(A_1) + S(A_2) + \cdots + S(A_m).$$

Then at least two of the sets $A_1, \ldots, A_m$ have a common interior point.

Suggestion. Assume, to the contrary, that the sets have no common interior points. Then

$$S(A_1 \cup A_2 \cup \cdots \cup A_m) = S(A_1) + S(A_2) + \cdots + S(A_m).$$

Since $A_1, \ldots, A_m \subseteq A$ and the complement $A - (A_1 \cup A_2 \cup \cdots \cup A_m)$ are measurable, we have

$$S(A_1 \cup A_2 \cup \cdots \cup A_m) \leq S(A),$$

a contradiction.

7.5.4. Let A be a measurable set, and let $A_1, \ldots, A_m$ be measurable subsets of A. Suppose that

$$nS(A) < S(A_1) + S(A_2) + \cdots + S(A_m)$$

for some positive integer $n < m$. Then at least $n + 1$ of $A_1, \ldots, A_m$ have a common interior point.

Suggestion. If no $n + 1$ sets share an interior point, then each interior point of the set $A_1 \cup \cdots \cup A_m$ is "counted" no more than n times in the sum

$$S(A_1) + S(A_2) + \cdots + S(A_m),$$

and therefore

$$S(A_1) + S(A_2) + \cdots + S(A_m) \leq nS(A).$$

7.5.5. A unit square contains a set whose area is more than $\frac{1}{2}$. Prove that this set contains two points, symmetric about the center of the square.

7.5.6. The area of a set on the sphere is greater than half of the area of the sphere. Prove that this set covers a pair of diametrically opposite points on the sphere.

The theorems of Blichfeldt and Minkowski

Fix a rectangular Cartesian coordinate system on the plane and through each point with integer coordinates draw two lines, parallel to the coordinate axes. The resulting system of lines is called an *integer lattice*, and points with integer coordinates are called *lattice points*. The integer lattice cuts the plane into unit squares.

Consider an integer lattice and a measurable plane set. The number of lattice points covered by the set depends not only on the shape of the set, but also its location. For example, there are sets with arbitrarily large area that do not cover a single lattice point (give an example!).

7.5.7. Blichfeldt's theorem. Let A be a measurable set on the coordinate plane with area greater than n. Then A can be translated so that it covers at least $n + 1$ lattice points.

Suggestion. The integer lattice cuts A into a finite number of pieces (the figure A is bounded!). The condition $S(A) > n$ shows that the number of pieces is not less than $n + 1$. Place all the squares that our figure intersects "in a single deck" (see Fig. 10). We will get at least $n + 1$ shapes inside a unit square with the total area greater than n.

Applying problem 7.5.4 to the unit square we see that there is a point P that belongs to at least $n + 1$ pieces of our set. It suffices to translate A by the vector that connects P with the lattice point.

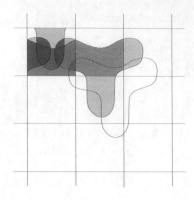

FIGURE 10

For $n = 1$, Blichfeldt's theorem can be reformulated as follows:

7.5.8. Let A be a measurable figure on the coordinate plane whose area is more than 1. Then A contains two distinct points (x_1, y_1) and (x_2, y_2) such that $x_2 - x_1$ and $y_2 - y_1$ are integers.

Recall that a plane figure A is called *convex* if the segment joining any two of its points lies entirely in A.

The following theorem, due to the German mathematician Hermann Minkowski, appears in geometric number theory.

7.5.9. Minkowski's theorem. Let A be a convex measurable set with area greater than 4 that is symmetric with respect to the origin. Then A contains a point with integer coordinates different from the origin.

Suggestion. Apply a homothety with the center at the origin and coefficient $\frac{1}{2}$ to A, obtaining the set B, whose area is greater than 1. By Blichfeldt's theorem, B contains distinct points (x_1, y_1) and (x_2, y_2), for which $x_2 - x_1$ and $y_2 - y_1$ are integers. By symmetry, $(-x_1, -y_1)$ also lies in B, and because B is convex, the midpoint O of the segment connecting $(-x_1, -y_1)$ and (x_2, y_2) also lies in B. The point O has the coordinates $\left(\frac{x_2 - x_1}{2}, \frac{y_2 - y_1}{2} \right)$. Therefore, the point with coordinates $(x_2 - x_1, y_2 - y_1)$ lies in A.

7.5.10. Show by an example that the condition $S(A) > 4$ in Minkowski's theorem cannot be replaced by $S(A) \geq 4$.

7.5.11. Let A be a measurable set on the coordinate plane whose area is less than n. Prove that A can be translated so that it covers at most $n - 1$ lattice points.

7.5.12. Let A be a convex measurable set that is symmetric with respect to the origin and has area greater than $4n$. Prove that A contains at least $2n + 1$ lattice points.

Dirichlet's theorem on approximation of irrational numbers

7.5.13. Dirichlet's theorem. For an arbitrary irrational number α and an arbitrary natural number s there exist integers x and y such that $0 < x \le s$ and

$$|\alpha x - y| < \frac{1}{s}.$$

Suggestion. We give a sketch of a proof using Minkowski's theorem. A direct proof can be obtained by following the suggestion to problem 7.5.14.
    Consider

$$A = \left\{ (x, y) \colon |\alpha x - y| < \frac{1}{s}, \quad |x| \le s + \frac{1}{2} \right\}.$$

This set is a parallelogram whose area is

$$\frac{2}{s} \cdot 2 \left(s + \frac{1}{2} \right) = 4 \left(1 + \frac{1}{2s} \right) > 4.$$

This figure is convex and symmetrical with respect to the origin (see Fig. 11). Minkowski's theorem states that in A there is a point with integer coordinates other than $(0, 0)$. We can assume that the first coordinate of this point is positive (explain this!). Thus, the theorem is proved.

7.5.14. Prove that for arbitrary irrational number α and natural number s there is a rational number $\frac{m}{n}$ such that $0 < n \le s$ and $\left| \alpha - \frac{m}{n} \right| < \frac{1}{ns}$.

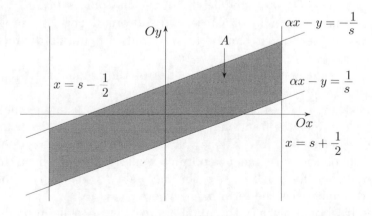

FIGURE 11

7.5.15. Prove that for an arbitrary irrational number α there are infinitely many rational numbers $\frac{m}{n}$ such that

$$\left| \alpha - \frac{m}{n} \right| < \frac{1}{n^2}.$$

Suggestions, solutions, and answers

7.5.1. Given a disk, one can circumscribe a regular n-gon around it and also inscribe a regular n-gon in it. The difference in the areas of these two polygons can be made arbitrarily small by choosing a sufficiently large n. Consequently, disks and segments of disks are measurable.

7.5.2. Any simple set is obviously bounded. Since a measurable set is contained in a simple set, it is also bounded.

7.5.5. Let F be the given figure, and let F' be the figure symmetric to it with respect to the center of the square. Then $S(F) + S(F') > 1$ and by the pigeonhole principle for areas (problem 7.5.3) there exists a point $X \in F \cap F'$. Obviously, X and the point X' symmetric to X form the required pair.

7.5.6. Consider a figure that is symmetrical to a given one relative to the center of the sphere, and repeat the arguments of the previous problem.

7.5.10. Consider the open square $\{(x, y) \colon |x| < 1, |y| < 1\}$.

7.5.11. Note that the half-open square $-k \le x, y < k$ covers exactly $4k^2$ lattice points when translated by any vector. Choose k large enough so that A is contained in some such half-open square K. By Blichfeldt's theorem, one can translate the set $K - A$ to cover at least $4k^2 - n + 1$ lattice points. Since all these lattice points lie in the translated image of the square K, the image of A will cover at most $n - 1$ nodes.

7.5.12. Apply a homothety with center at the origin and coefficient $\frac{1}{2}$ to A, getting the figure B whose area is greater than n. It follows from Blichfeldt's theorem that B contains distinct points $(x_0, y_0), \ldots, (x_n, y_n)$, for which all the differences $x_i - x_j$ and $y_i - y_j$ are integers. We can assume that $x_0 \ge x_1 \ge \cdots \ge x_n$ and that among the points (x_i, y_i) for which $x_i = x_0$, the maximum value of the second coordinate is y_0. As in the proof of Minkowski's theorem, it can be shown that A contains distinct points $(0, 0)$, $(x_0 - x_i, y_0 - y_i)$, $(x_i - x_0, y_i - y_0)$.

7.5.14. It suffices to divide both sides of the inequality in Dirichlet's theorem by x.

Alternatively, you can solve this problem without using geometric considerations: consider the fractional parts of the numbers $\alpha, 2\alpha, \ldots, s\alpha$ and divide the segment $[0, 1]$ into s equal parts. There are two cases possible.

1. Each of the s segments contains exactly one of the numbers α, $2\alpha, \ldots, s\alpha$. Then for some $n \le s$ the inequality $\{n\alpha\} < 1/s$ holds, and the desired number has the form m/n, where $m = [n\alpha]$.

2. The fractional parts of the numbers $n_1\alpha$ and $n_2\alpha$ lie in one segment. Then the desired number is m/n, where $m = |[n_1\alpha] - [n_2\alpha]|$, $n = |n_1 - n_2|$.

6. Phase spaces (3) *By A. Ya. Kanel-Belov*

Informally, the phase space of a system is the set of all its possible states.

For example, the phase space for a system where a state is a real number is a straight line. The phase space of the system of ordered pairs of real numbers is the plane. If the coordinates are restricted to intervals, the phase state is a rectangle.

The idea of phase space is indispensable for defining geometric probabilities (see Section 2 of Chapter 6 and [**Vas91**]).

7.6.1. Konstantinov's cart problem. Two nonintersecting (except at endpoints) roads lead from A to B. It is known that cars (points) connected by a rope of length shorter than $2L$ are able to travel from A to B along the two different roads without breaking the rope. Will round carts of radius L be able to pass if they travel towards each other along these two roads?

7.6.2. Decanting. Three containers have volumes of 6, 7, and 12 liters, respectively. The two smaller containers are filled. Is it possible to pour 9 liters of liquid into the largest container?

7.6.3. External "billiards." (a) Consider a square on a plane with a point A_1 lying outside the square at a distance a from the center of the square. From this point a line is drawn, intersecting the square at a single point B_1. The starting point A_1 is reflected symmetrically with respect to the point B_1, obtaining the point A_2. From A_2 we again draw a new line, intersecting the square at the single point $B_2 \neq B_1$ (assuming that such a line exists). Reflect A_2 with respect to B_2 to obtain A_3, and so on. Suppose that at each step it is possible to draw a straight line to continue this process. Is it possible that eventually the distance from A_n to the center of the square exceeds $1\,000\,000 \cdot a$?

(b)* What will happen if instead of a square we consider an arbitrary polygon?

7.6.4. Three pedestrians walk along straight roads on a plane, each at their own constant speed. Prove that if their locations were collinear at any three time moments, then the locations are always collinear.

7.6.5. In Moscow there are 7 high-rise buildings. A visiting mathematician wants to find a point from which they are all visible in a given order (starting with Moscow State University, going clockwise). Is this possible for any given order?

7.6.6. N lines are marked in the plane, where no three intersect at one point, no two are parallel, and any three of their intersection points either do not lie on a straight line or lie on one of the marked lines. Call a line in the plane *good* if it does not pass through the intersection points of marked lines. In how many different ways can one add a good line? We consider two methods the same if one can be obtained from the other by continuously moving a new straight line so that it remains good all the time.

7.6.7. On the unit sphere there is a curve (consisting of several arcs of circles) of length less than π. Prove that there is a great circle that does not intersect this curve.

7.6.8. On the unit sphere lies a curve (consisting of several arcs of circles) of length greater than $k\pi$. Prove that there is a great circle that crosses this curve at least at k points.

7.6.9. We are given a polygon of unit area and also 1000 points on the plane. Prove that you can move the polygon by a vector of length less than $\sqrt{\frac{1000}{\pi}}$ so that it does not cover any of the points.

7.6.10. In a unit square there is a piecewise linear path of length greater than 1000. Prove that there is a horizontal or vertical line which intersects this path at least at 500 points.

7. Linear variation (3)
By A. Ya. Kanel-Belov

The notion of linearity is central to the problems in this section.

A *linear function* is a function of the form $f(x) = ax + b$, where a and b are real numbers.

Important properties of linear functions:

• a linear function either always increases or always decreases or remains constant;

• a linear function reachs its maximal (minimal) values at the endpoints of an interval.

7.7.1. The area of a triangle with vertices on the boundary of a parallelogram does not exceed half the area of the parallelogram.

7.7.2. Prove that the area of the pentagon $ABCDE$ is less than the sum of the areas of the triangles ABC, BCD, CDE, DEA, EAB.

7.7.3. A circle contains n nonoverlapping triangles. What is the maximal value of the ratio of the total area of the triangles to the area of the circle?

7.7.4. Let $0 < a_0 < a_1 < \cdots < a_n$. Prove that the expression

$$a_0 + a_1 \cos x + \cdots + a_n \cos nx$$

(called a *trigonometric polynomial*) is equal to 0 for exactly n values of x on the interval $[0, \pi]$.

7.7.5. A crazy architect has n identical bricks. He wants to build an n-level tower that will extend horizontally as far as possible. Each tower level consists of exactly one horizontal brick. For each $k = 1, \ldots, n$, the projection of the center of gravity of the bricks of the tower that are located above the kth level should fall on the kth-level brick. What will be the maximum possible horizontal length of the entire tower?

FIGURE 12

7.7.6. Two travelers decided to visit all the cities of Russia. They started independently of each other and, possibly, from different cities. The first traveler, upon leaving one city, always goes to the farthest city which he has not yet visited. The second traveler always chooses the nearest city. Prove that after visiting all the cities, the route of the first traveler is not shorter than the route of the second.

7.7.7. Define a *rounding* of a number to be the replacement of this number by some integer (not necessarily the closest one). Given n numbers, prove that they can be rounded so that the sum of any m of them differs from the sum of their roundings by not more than $(n+1)/4$.

7.7.8. The numbers $x_1, x_2, \ldots, x_n$ lie between 0 and 1 (possibly on an endpoint). Find the maximal possible value of the expression

$$\sum_{1 \leq i < j \leq n} |x_i - x_j|.$$

7.7.9. On the unit circle, n points $a_1, \ldots, a_n$ are marked in counterclockwise order so that the polygon formed by them contains the center of the circle.

(a) Prove that there are two points at a distance at least $4/n$ between them.

(b) Prove that there is a point located at a distance at least $8/n^2$ from the midpoint of the segment connecting its neighbors.

7.7.10. Several points are marked on the plane, not all of them lying on the same line. To each point a real number is assigned. Define the *increment* of an arbitrary line containing at least two marked points to be the sum of all the numbers assigned to the marked points on this line. Suppose that the increment of every line is equal to 0. Prove that all the assigned numbers are equal to 0.

7.7.11. Suppose that $a_1, \ldots, a_k$ are such that for all x, the following inequality holds: $a_1 \cos x + a_2 \cos 2x + \cdots + a_k \cos kx \geq -1$. Prove that $a_1 + a_2 + \cdots + a_k \leq k$.

7.7.12. There are 300 boots in the warehouse: 100 rubber, 100 canvas, 100 wool. Among those 300 boots, there are an equal number of left and right boots. Prove that we can make 50 correct pairs (i.e., for which the right and left boots are of the same material).

In conclusion, we suggest that you think about problems 7.8.9 and 7.8.10 in Section 8 of this chapter and consider the following difficult problem.

7.7.13.* The plane contains $n > 2$ nonparallel lines, not all of which go through one point. Prove that among the polygons into which they divide the plane, you can find $n - 2$ triangles (with no lines in the interior).

Suggestions, solutions, and answers

7.7.2 and 7.7.13. These problems are discussed in detail in [**K-BK92**].

8. Compose a square (3*) *By M. B. Skopenkov, O. A. Malinovskaya, S. A. Dorichenko, and F. A. Sharov*

This section is devoted to solving the following problem (in special cases):

Problem. When can a square be composed using rectangles similar to a given one?

While solving this problem, we will see beautiful applications of algebra. Namely, from systems of linear equations and polynomials with integer coefficients we come to combinatorial geometry. You will need to have some knowledge of these topics (see, for example, [**G**]). It is also desirable to know something about cutting problems; see, for example, [**S87**].

Our approach to solving these problems develops some ideas from the book [**Yag68**].

Another approach to solving relies on a physical interpretation using electrical circuits (although it is easier to not use this method). You can learn about this physical interpretation and its application to the solution

of the problem posed in [**SPD**, **SMD**]. A fascinating story about the history
of this approach can be found in the book [**G**].

Leading questions

> "I have a thought!," said Boa Constrictor, opening
> its eyes. "A thought. And I am thinking it."
> "What is a thought?," asked Monkey.
> "It's hard to say so right away...."
> "Wow!," Monkey jumped up. "Oh, what a good
> thought. And may I also think it a little?"
>
> *G. Oster*, Grandma Boa

7.8.1.° Is it true that for any positive integers m and n you can compose a
square out of $m \times n$ rectangles?

7.8.2. A designer ordered frames for a square window. The projects (see
Fig. 13(A) and (B)) show how the glass should adjoin together and how
they should be oriented (short or long side up). Is it possible to make all
the glass in these two square windows to be similar rectangles?

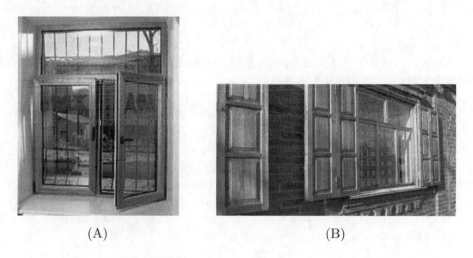

(A)   (B)

FIGURE 13. Window frame designs; see problem 7.8.2.

7.8.3. Is it possible to cut a square into three similar but noncongruent
rectangles?

7.8.4. Is it possible to cut a square into 5 squares?

7.8.5. All the shelves of the cabinet in Fig. 14(C), as well as all the scraps
from which the quilt is sewn in Fig. 14(D) are squares. Are the cabinet and
the quilt themselves square?

(C)

(D)

FIGURE 14. Cabinet and quilt; see problem 7.8.5.

7.8.6. Is it possible to tile the plane with pairwise distinct squares with integer side lengths?

7.8.7. Is it possible to cut a square into rectangles with aspect ratio $2+\sqrt{2}$? Same question for $2-\sqrt{2}$, for $3+2\sqrt{2}$, and for $3-2\sqrt{2}$.

7.8.8. Can $1+\sqrt{2}$ be expressed as the sum of squares of numbers of the form $a+b\sqrt{2}$, where a and b are rational?

Definition. Consider a rectangular sheet of paper with a cutting into rectangles drawn on it. One can cut the rectangle along any line into two rectangles, then perform this operation with each piece separately, and so on. Such cuttings are called *trivial*. For example, the cuttings in Fig. 13 are trivial, and those in Fig. 14 are nontrivial.

   In the next four problems, first assume that the cuttings are trivial, and then think about arbitrary cuttings. Hints for difficult problems will be given.

7.8.9. What rectangles can be cut (trivially) into rectangles with one side of length 1?

7.8.10. What rectangles can be cut (trivially) into squares?

7.8.11. Is it possible to cut a square (trivially) into rectangles with aspect ratio $\sqrt{2}$? Same question for $1+\sqrt{2}$.

By *good* numbers we mean all numbers that can be represented as $x = a + b\sqrt{2}$ with rational a and b.

7.8.12 (The main problem)**.** For what good x can a square be cut (trivially) into rectangles with aspect ratio x?

Rectangles from squares.

> You road I enter upon and look around,
> I believe you are not all that is here,
> I believe that much unseen is also here.
> *Walt Whitman*, Song of the Open Road

In this subsection, we outline a new version of the elementary solution to problems 7.8.10 and 7.8.12. Below, the letters a, b, c, d (sometimes with indices) denote *rational* numbers.

7.8.13. Can a $1 \times \sqrt{2}$ rectangle be cut into squares with rational sides? With sides that are either rational or have the form $b\sqrt{2}$? With sides that are arbitrary good numbers? The same questions for a $1 \times (1+\sqrt{2})$ rectangle and for a $1 \times (2 + \sqrt{2})$ rectangle.

To prove the impossibility of cutting, it is natural to consider the area and its *additivity*: the area of the whole is equal to the sum of the areas of the parts. It is unlikely that it will be possible to answer the question in problem 7.8.13 for the $1 \times (2 + \sqrt{2})$ rectangle without the following generalization of the concept of area (we generalize the concept of area so that the area of this rectangle becomes negative and area of squares remains nonnegative).

Definition. Let x be a real number. The *x-area* (also called the *Hamel area*) of an $(a + b\sqrt{2}) \times (c + d\sqrt{2})$ rectangle is defined to be $(a + bx)(c + dx)$. The number $\bar{s} := a - b\sqrt{2}$ is called *conjugate* to $s = a + b\sqrt{2}$.

7.8.14. The usual area of a $(a + b\sqrt{2}) \times (c + d\sqrt{2})$ rectangle and the number conjugate to it are possible x-areas of the rectangle. What is x in each case?

7.8.15. Find all rectangles of the form $(a + b\sqrt{2}) \times (c + d\sqrt{2})$ whose x-area is nonnegative for all x.

7.8.16. Additivity of the x-area. If a rectangle is cut into a finite number of rectangles whose sides are good numbers, then for any $x \in \mathbb{R}$, the x-area of the original rectangle is the sum of x-areas of the rectangles into which it is cut.

Suggestion. Start with the case of a rectangle cut into two rectangles.

7.8.17. Solve problems 7.8.10 and 7.8.12 for the special case where the sides of all the squares and all the rectangles involved in the construction are good numbers (cutting is not necessarily trivial).

To solve problems 7.8.10 and 7.8.12 in the general case, the definition of x-area is no longer helpful; after all, it is defined only for good numbers, and now squares with any side may occur.

In the next three problems, we suppose that the rectangle $s_0 \times t_0$ is cut into the rectangles $s_1 \times t_1$, $s_2 \times t_2$, ..., $s_N \times t_N$, and s_0 and t_0 are incommensurable (i.e., the ratio s_0/t_0 is irrational).

7.8.18. Denote
$$P = \{s_0, t_0, s_1, t_1, \ldots, s_N, t_N\}.$$
Then you can choose numbers $e_1, e_2, \ldots, e_n \in P$, so that any number $p \in P$ is uniquely represented as
$$p = as_0 + bt_0 + a_1e_1 + a_2e_2 + \cdots + a_ne_n.$$

Suggestion. Begin with the example shown in Fig. 15.

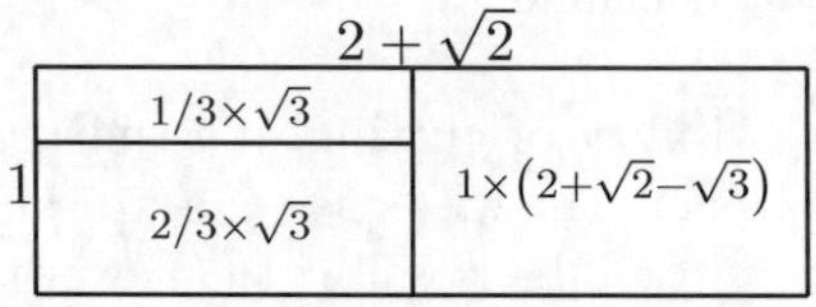

FIGURE 15. Toward constructing a basis.

Fix a set of numbers $s_0, t_0, e_1, e_2, \ldots, e_n$ from problem 7.8.18 above and call it a *basis*.

Definition. Let y be a real number. We define the *y-area* of a rectangle with sides
$$as_0 + bt_0 + a_1e_1 + a_2e_2 + \cdots + a_ne_n$$
and
$$cs_0 + dt_0 + c_1e_1 + c_2e_2 + \cdots + c_ne_n$$
to be $(a + by)(c + dy)$.

Note that for $y = x$ and good incommensurable s_0, t_0, this definition is not always equivalent to the definition of the x-area given earlier!

7.8.19. Calculate the y-area of the initial $s_0 \times t_0$ rectangle. Is it nonnegative for all y?

7.8.20. Prove that for any y, the y-area of the initial $s_0 \times t_0$ rectangle is equal to the sum of the y-areas of the rectangles into which it is cut.

7.8.21. Dehn's theorem. If a rectangle can be cut into squares (not necessarily congruent), then the ratio of its sides is rational.

7.8.22. If the 1×1 square is cut into rectangles such that the aspect ratio of each of them is a good number, then the lengths of sides of all rectangles are good numbers.

From cutting to roots of polynomials

> Here is the test of wisdom,
> Wisdom is not finally tested in schools. . . .
> *Walt Whitman*, Song of the Open Road

7.8.23. A rectangle is cut into finitely many rectangles with aspect ratio r. Prove that the aspect ratio of the large rectangle is $P(r) : Q(r)$, where $P(x)$ and $Q(x)$ are polynomials with integer coefficients.

7.8.24. These polynomials can be chosen so that

$$P(-x)/Q(-x) = -P(x)/Q(x) \quad \text{for all } x$$

and

$$P(x)/Q(x) > 0 \quad \text{for all } x > 0.$$

7.8.25. From several rectangles with aspect ratio r a square is assembled. Prove that r is a root of a nonzero polynomial with integer coefficients.

What's next

> *FAUST:*
> Wohin soll es nun gehn?
> *MEPHISTOPHELES:*
> Wohin es dir gefällt.
> Wir sehn die kleine, dann die große Welt.
> Mit welcher Freude, welchem Nutzen
> Wirst du den Cursum durchschmarutzen!
> *Goethe*, Faust

> *Faust:*
> Where will we go, then?
> *Mephistopheles:*
> Where you please.
> The little world, and then the great, we'll see.
> With what profit and delight,
> This term, you'll be a parasite!
> *Goethe*, Faust

But what can be said in the case of arbitrary r, not necessarily a good one? We give the answer here, without proof (for a proof see [**PraSko**]).

Freiling–Laczkovich–Rinne–Szekeres theorem (1994). *For $r > 0$, the following three conditions are equivalent.*
  1. *A square can be cut into rectangles with the aspect ratio r.*
  2. *For some positive rational numbers c_i the following equality holds:*

$$c_1 r + \cfrac{1}{c_2 r + \cfrac{1}{c_3 r + \cdots + \cfrac{1}{c_n r}}} = 1.$$

  3. *The number r is the root of a nonzero polynomial with integer coefficients such that all its complex roots have a positive real part.*

Suggestions, solutions, and answers

7.8.1. Of course, if the ratio of the sides of the rectangle is $m : n$, where m and n are integers, then a square can be constructed with mn such rectangles; see Fig. 17. In our figure, the rectangles are congruent and are arranged side by side. It is clear that with this method we will be able to construct the square only if the aspect ratio is rational.

7.8.2. *Answer in case* (A): impossible.

Solution. Assume that three similar rectangles are located as in Fig. 18; the top one is horizontal, and the two remaining are vertical.

Since the bottom rectangles are arranged vertically, their width is equal to half the side of the square. Their length is more than half, but less than the whole side length of the square. The width is obviously *more* than half the length.

The width of the top rectangle and the length of the bottom one together give the length of a side of the square. Therefore, the width of the top rectangle is obviously *less* than half the length. Therefore these rectangles cannot be similar.

Answer in case (B): possible.

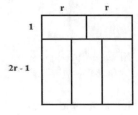

FIGURE 16. Cutting into 5 rectangles.

Solution (from [**SMD**]). Consider the partition into 5 similar rectangles in Fig. 16 where the two top ones are horizontal and the remaining three are vertical. Since the top rectangles are similar and lie horizontally, they are equal. Let their sizes be $1 \times r$; then the side of the square is $2r$. Since the bottom rectangles are similar to the top ones and are arranged vertically, we obtain the equation $3\frac{2r-1}{r} = 2r$. Solving it, we find the two roots $\frac{3\pm\sqrt{3}}{2}$. For each of these roots, it is easy to construct the desired cutting.

FIGURE 17   FIGURE 18

7.8.3. *Answer*: yes.

Solution. Consider a $k \times 1$ rectangle, where we will specify k later. We will call this rectangle *standard*. Attach a $k \times k^2$ rectangle to it, as shown in Fig. 19. Note that this rectangle is similar to the standard one with similarity coefficient k. The resulting rectangle has a side of length $1 + k^2$. Attach a rectangle similar to the standard one along the top side. Then the lateral side has length $k(1+k^2) = k^3+k$. The resulting rectangle is a square if $1 + k^2 = k^3 + 2k$.

Consider the difference $f(k)$ between the length of two sides of the resulting rectangle; i.e., $f(k) = k^3 - k^2 + 2k - 1$. For $k = 0$ it is equal to -1, and for $k = 1$ it is equal to $+1$. By the intermediate value theorem, $f(k)$ vanishes at some point in the interval $(0,1)$. This is the required number k.

Now it is easy to prove the existence of the desired cutting. The side lengths of all rectangles are already expressed in terms of k, and from the equality $f(k) = 0$ it follows that the resulting rectangle is a square. No rectangles in the cut are congruent, since their largest sides are not equal: $k < 1 < 1 + k^2$.

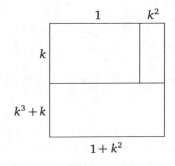

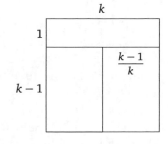

FIGURE 19   FIGURE 20

7.8.4. *Answer*: no (this is a problem from the Moscow Mathematical Olympiad).

Suggestion. See Problem 3 in [**SPD**].

Solution. It can be foud in the book [**Yag68**].

7.8.5. *Answers*: C: no; D: yes.

Suggestion for C (from [**SPD**]). Enumerate the squares (shelves) as shown in Fig. 21. We assume that the horizontal side of the rectangle (cabinet) is 1, and we denote the vertical side (excluding legs) by x. The side of square number k is denoted by x_k.

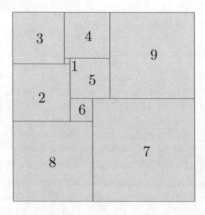

FIGURE 21. Cutting a rectangle into 9 unequal squares.

The left side of the rectangle is made of sides of squares 2, 3, and 8, whence $x = x_2 + x_3 + x_8$. Squares 1 and 4 are adjacent to the right side of square 3, so $x_3 = x_1 + x_4$. Similarly, $x_6 + x_8 = x_7$ and $x_1 + x_2 = x_5 + x_6$, $x_4 + x_5 = x_9$. We do not write down the condition for the right side of the rectangle since it follows from the previous ones (obtained by adding all the equalities).

Considering the horizontal sides, we get $1 = x_3 + x_4 + x_9$, $x_4 = x_1 + x_5$, $x_1 + x_3 = x_2$, $x_5 + x_9 = x_6 + x_7$, $x_2 + x_6 = x_8$.

Solving the resulting system of linear equations (for a solution method, see [**G**]), we find

$$x = 33/32, \quad x_3 = 9/32, \quad x_7 = 9/16, \quad x_1 = 1/32, \quad x_4 = 1/4,$$
$$x_9 = 15/32, \quad x_5 = 7/32, \quad x_2 = 5/16, \quad x_8 = 7/16, \quad x_6 = 1/8.$$

In particular, $x \neq 1$. In our example, the rectangle was cut into distinct squares.

7.8.6. *Answer*: yes.

Solution (from [**SPD**, Solution of Problem 1]). Take the example shown in Fig. 21, where the sides of all squares are integer; for example, you can take a partition in which a 33×32 rectangle is cut into squares (the length of the vertical side is indicated first, then the horizontal one). We can tile the plane with this construction. Place a 33×33 square on the left of our rectangle, then place a 65×65 square on the resulting 33×65 rectangle, then place a 98×98 square on the 98×65 rectangle, and then we place a 163×163 square on the bottom, and so on. As a result, we tile the plane "with a spiral," and each new square has a side equal to the larger side of the rectangle to which it is attached. Therefore, it is larger in size than all previous squares.

7.8.7. *Answer*: it is possible.

Solution. See Fig. 22.

Path to solution. Consider a rectangle with sides of length 1 and $a+b\sqrt{2}$. Rectangles of size $k \times k(a + b\sqrt{2})$ have the same aspect ratio. If for given a and b we can find two rectangles of this kind with the additional condition that in each of them one side has integer length and the sum of the two remaining lengths is an integer, then the problem is solved. From rectangles with integer sides, you can assemble a square. To answer the first question of the problem, we take rectangles of sizes $1 \times (2 + \sqrt{2})$ and $(2 - \sqrt{2}) \times 2$ and assemble a square as shown in Fig. 22(A). Similarly, for the remaining questions, rectangles with sides 1, $2 - \sqrt{2}$, and $2 + \sqrt{2}$, 2 work (see Fig. 22(B)) and with sides $1 \times (3 + 2\sqrt{2})$ and $(3 - 2\sqrt{2}) \times 1$. The partitions in the last two cases coincide (see Fig. 22(C)).

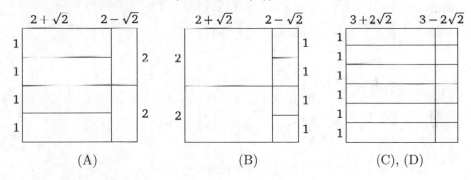

FIGURE 22

7.8.8. *Answer*: no.

Hint. If the number is a sum of squares, then the conjugate to it also has this property.

7.8.9. *Answer*: rectangles for which one of the sides has integer length (*De Bruijn's theorem*).

Solution. See [**TF**, Theorem 4 on p. 25].

7.8.10. *Answer*: rectangles with a rational aspect ratio (*Dehn's theorem*).

Suggestion. One path to the proof was outlined in problems 7.8.13–7.8.21.

Solution for the case of trivial partitions (written by A. Balakin). Let us prove that if a rectangle can be trivially cut into squares, then the ratio of its sides is rational.

Take an arbitrary trivial cutting of a rectangle into squares. First, we divided our rectangle into two. At the next step, we divided each of the newly formed pieces into two, or leave it as is, etc. At the end, we obtained a set of squares, into which our original rectangle was cut.

Now we take this set of squares and begin to perform the inverse process; i.e., we glue at each step a certain number of pairs of rectangles along the side so that at the end the original one is obtained.

Let us prove that if two rectangles with rational aspect ratios are glued along the side, then the aspect ratio of the resulting rectangle is also rational.

Indeed, suppose we glue them along the side of length a. Then, for one of them, the lengths of the sides are a and pa, and for the second one they are a and qa, where p and q are rational. The glued rectangle will have sides a and $pa + qa$, and therefore, the rational aspect ratio is $p + q$.

The squares we start with have the aspect ratio 1. This is a rational number. Therefore, each of the obtained rectangles and, therefore, the original one, has a rational aspect ratio.

7.8.11. *Answer*: for $\sqrt{2}$ the answer is no, and for $1 + \sqrt{2}$ the answer is no.

Hint for $\sqrt{2}$. Reduce to problem 7.8.10 using a scaling in one direction with the coefficient $1/\sqrt{2}$; see [**SPD**, solution of problem 7].

Solution for $1 + \sqrt{2}$. It follows from problem 7.8.12.

7.8.12. *Answer*: for $x = a + b\sqrt{2}$ such that $a^2 > 2b^2$, where a, b are rational numbers.

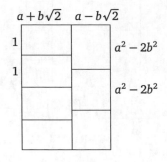

FIGURE 23

Suggestion. See Fig. 23. Use problems 7.8.22 and 7.8.17. For a more general discussion, see [**Sh**].

For a solution using a physical interpretation, see [**SMD**] (but the non-physics solution is easier).

7.8.13. *Answer*: no (for all questions).

Solution (L. Aliyeva). For the first question, suppose that a $1 \times \sqrt{2}$ rectangle is cut into squares with rational sides. The sum of the lengths of the sides of the squares adjacent to the side of the rectangle with length of $\sqrt{2}$ should be equal to $\sqrt{2}$. Therefore, such a cutting is impossible, since the sum of rational numbers cannot be irrational.

Now we answer the second question. Suppose that the $1 \times \sqrt{2}$ rectangle is cut into squares with sides of the form a or $b\sqrt{2}$. The area of the rectangle is $\sqrt{2}$. The areas of the squares are either a^2 or $2b^2$. The area of the rectangle is equal to the sum of the areas of squares. We get a contradiction since an irrational number cannot be equal to the sum of rational ones. Here we used the additivity of the area.

Answering the third question, suppose that a $1 \times \sqrt{2}$ rectangle is cut into n squares with sides $a_1 + b_1\sqrt{2}$, $a_2 + b_2\sqrt{2}$, ..., $a_n + b_n\sqrt{2}$. The areas of the squares are $a_1^2 + 2a_1b_1\sqrt{2} + 2b_1^2$, ..., $a_n^2 + 2a_nb_n\sqrt{2} + 2b_n^2$, respectively. Denote $A := a_1^2 + \cdots + a_n^2$, $B := b_1^2 + \cdots + b_n^2$, $C := a_1b_1 + \cdots + a_nb_n$. Then A, B, C are rational numbers. The area of the rectangle is the sum of the areas of the squares, so $\sqrt{2} = A + 2C\sqrt{2} + 2B$. Thus $\sqrt{2} \cdot (1 - 2C) = A + 2B$. If $C = 1/2$, then $A = B = 0$ and $a_1 = \cdots = a_n = b_1 = \cdots = b_n = 0$, which is impossible. Therefore, both sides of the equation can be divided by $1 - 2C$. Thus, $\sqrt{2} = (A + 2B)/(1 - 2C)$. We get a contradiction since an irrational number cannot be equal to the quotient of two rational ones. Here we again used the additivity of the area.

The answer to the fourth question follows from problem 7.8.8, and the answer to the fifth question follows from problem 7.8.17.

The proposed method for proving the impossibility of a cutting is to find an x such that the x-area of the cut figure is negative, but the x-area of all figures into which we cut it is nonnegative. Then the additivity of x-area (problem 7.8.16) produces a contradiction, so we conclude that the cutting is impossible.

7.8.16. It is easy to verify that the sum of the x-area of *two* rectangles with good sides having a common side is equal to the x-area of their union. Indeed, let a rectangle with x-area S consist of two rectangles with x-areas S_1 and S_2 (see Fig. 24, left). Then

$$S_1 + S_2 = (a + bx)(c_1 + d_1x) + (a + bx)(c_2 + d_2x)$$
$$= (a + bx)((c_1 + c_2) + (d_1 + d_2)x) = S.$$

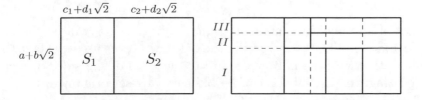

FIGURE 24. Proof of additivity; see solution to problem 7.8.16.

Now suppose that the number of rectangles in the cutting is greater than two. Extend each cut as shown in Fig. 24 on the right. Then each new rectangle will also have good side lengths. Consider horizontal layers of rectangles sequentially attached to each other along the common side of rectangles (I, II, III in Fig. 24 on the right). Using the already proved additivity of x-area for two rectangles with a common side, it is easy to prove by induction that the x-area of any such layer is equal to the sum of x-areas of the rectangles in this layer. Now we attach these layers to each other and use the additivity of x-areas of a series of rectangles. We conclude that the x-area of the original rectangle is equal to the sum of x-areas of the horizontal layers. This sum is equal to the sum of the x-areas of all the rectangles in the cutting.

7.8.17. In problem 7.8.10 it suffices to prove the impossibility of cutting rectangles of the form $1 \times (c + d\sqrt{2})$, $d \neq 0$, into squares. Let such a rectangle be cut into squares with good sides. Its x-area is $c + dx$. Since $d \neq 0$, this x-area is negative for some x. At the same time, the x-area of any square with side $a + b\sqrt{2}$ is $(a + bx)^2$, which is nonnegative for any x. We get a contradiction since the sum of nonnegative numbers cannot be negative.

In problem 7.8.12, consider a number conjugate to the ordinary area.

7.8.18. We write consecutively the lengths of the sides of the rectangles in the cutting, starting with s_0 and t_0. For example, for the cutting in Fig. 15 we get the following sequence:

$$s_0 = 1, \qquad t_0 = 2 + \sqrt{2}, \qquad s_1 = 1/3, \qquad t_1 = \sqrt{3},$$
$$s_2 = 2/3, \qquad t_2 = \sqrt{3}, \qquad s_3 = 1, \qquad t_3 = 2 + \sqrt{2} - \sqrt{3}.$$

Next, in this sequence we underline the numbers that cannot be represented as a linear combination of the previous ones with rational coefficients (we will explain what this means). For example, we underline s_0 (there are no previous numbers at all). We also underline the number t_0, since it is incommensurable with s_0 and, therefore, it cannot be represented in the form as_0. Then, if s_1 can represented in the form $as_0 + bt_0$, then s_1 is not underlined, but otherwise it is underlined. Similarly, if t_1 can be represented in the form $as_0 + bt_0 + a_1 s_1$, then t_1 is not underlined, and if it is not, it is underlined, and so on. In our example, we get the following sequence:

$$\underline{1}, \quad \underline{2 + \sqrt{2}}, \quad 1/3, \quad \underline{\sqrt{3}}, \quad 2/3, \quad \sqrt{3}, \quad 1, \quad 2 + \sqrt{2} - \sqrt{3}.$$

The set of all underlined numbers s_0, t_0, e_1, e_2, $\ldots$, e_n will be the basis.

Comment. When constructing the basis, we never used the fact that the numbers that we originally wrote are the lengths of the sides of rectangles in a cutting; that is, a basis can be built for any set of numbers.

7.8.19. *Answer:* y.

7.8.20. The proof repeats the solution to problem 7.8.16 with the only difference that the x-area should be replaced with the y-area and all numbers of the form $a + b\sqrt{2}$ should be replaced with the corresponding numbers of the form $as_0 + bt_0 + a_1 e_1 + a_2 e_2 + \cdots + a_n e_n$ (including those in Fig. 24).

7.8.21. Let an $s_0 \times t_0$ rectangle be cut into squares, with s_0 and t_0 being incommensurable. By definition, the y-area of this rectangle is y. Choose

$y < 0$. The side of any square in the cutting is of the form $as_0 + bt_0 + a_1 e_1 + a_2 e_2 + \cdots + a_n e_n$. The y-area of this square is $(a + by)^2$, which is nonnegative for *any* y. For $y < 0$, we have a contradiction: the sum of nonnegative numbers cannot be negative.

Comment. Note that y-area is not a trick, but an example of a "signed measure", which is an important concept in mathematics. For other proofs of Dehn's theorem, see [**SPD**] or [**Yag68**, § 3].

7.8.23 and 7.8.25. See article [**SMD**].

9. Is it possible to make a cube from a tetrahedron?[2] (3)
By M. V. Prasolov and M. B. Skopenkov

This section is devoted to the proof of the following statement.

Dehn's theorem (Hilbert's third problem). *A regular tetrahedron cannot be cut into finitely many convex polytopes that can be assembled into a cube.*

Surprisingly, one can prove this theorem by investigating only cuttings of rectangles, not polytopes! This section develops a new version of the elementary proof given in [**Fuks**] based on this idea. To solve these problems the reader should be familiar with Section 8.3 in [**ZS**] and Section 4 in Chapter 4 in the present book

7.9.1. (a) Cut a $1 \times 2 \times 4$ rectangular box into two convex polytopes that make up a cube.

(b) Cut a cube into 6 congruent pyramids.

(c) Cut a cube into 6 congruent tetrahedra.

(d)* Cut a $1 \times 2 \times 3$ rectangular box into several convex polytopes that make up a cube.

Reduction to a plane geometry problem

Let M be a convex polytope. Enumerate its edges $1, 2, \ldots, n$, and let $l_1, l_2, \ldots, l_n$ be the corresponding lengths and $\alpha_1, \alpha_2, \ldots, \alpha_n$ the corresponding dihedral angles. To M we associate a set of rectangles of size $l_i \times \alpha_i$ on the plane where l_i is the horizontal size and α_i is the vertical size (see Fig. 25).

Two such sets are called *rectangular scissor-congruent* if the rectangles of one set can be cut into smaller rectangles from which rectangles of the second set can be assembled using only parallel translations (see Fig. 26). We say that two polytopes are *scissor-congruent* if one of them can be cut

---

[2]The authors are grateful to S. Dorichenko, A. Zaslavsky, K. Kokhas, B. Frenkin, G. Chelnokov, and A. Shapovalov for useful discussions.

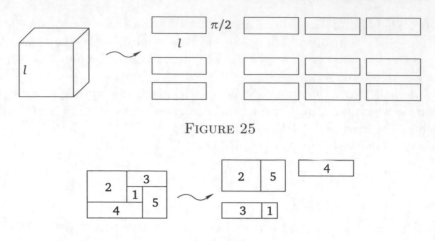

FIGURE 25

FIGURE 26

into several smaller convex polytopes, from which you can assemble the second polytope, rotating parts freely.

7.9.2.° Consider the set of rectangles corresponding to the regular tetrahedron with side a and dihedral angles θ between faces. To which of the following rectangles is this set rectangular scissor-congruent?
(1) $4a \times \theta$;   (2) $4a \times 4\theta$;   (3) $6a \times \theta$;   (4) $6a \times 6\theta$.

The following lemma reduces the question of scissor-congruence of polytopes to a two-dimensional problem.

Reduction lemma. *If two polytopes are scissor-congruent, then their corresponding sets of rectangles will be rectangular scissor-congruent after possibly adding rectangles of type $l \times \pi$ to these sets.*

The proof of this lemma is contained in problem 7.9.3.

7.9.3. Suppose that a convex polytope M is cut into several convex polytopes $M_1, M_2, \ldots, M_k$.
  (a) Let e be an edge of the polytope M, l the length of e, and α the dihedral angle at e. Let $l_1, l_2, \ldots, l_n$ be the lengths of all edges of the polytopes M_i lying on the edge e, and let $\alpha_1, \alpha_2, \ldots, \alpha_n$ denote the dihedral angles at these edges. Then the $l \times \alpha$ rectangle can be cut into the $l_1 \times \alpha_1, \ldots, l_n \times \alpha_n$ rectangles.
  (b) Let ℓ be a line in space that does not contain the edges of M. Let $l_1, l_2, \ldots, l_n$ be the lengths of all the edges of the polytopes M_i lying on ℓ, and let $\alpha_1, \alpha_2, \ldots, \alpha_n$ be the dihedral angles at the corresponding edges. Then the set of $l_1 \times \alpha_1, \ldots, l_n \times \alpha_n$ rectangles is rectangular scissor-congruent to an $l \times \pi$ rectangle.
  (c) Prove the reduction lemma.

To prove that the regular tetrahedron and cube of the same volume are not scissor-congruent, we will show that the corresponding sets of rectangles are not rectangular scissor-congruent. To do this, we need the following result.

(d) The dihedral angle θ of the edge of a regular tetrahedron is incommensurable with π; i.e., the ratio θ/π is irrational.

Thus, the proof of Dehn's theorem is reduced to the following statement.

Incommensurability lemma. *If θ and π are incommensurable (i.e., θ/π is irrational), then for any a and b, the $a \times \theta$ and $b \times \pi$ rectangles are not rectangular scissor-congruent. Moreover, they will remain not rectangular scissor-congruent after adding any rectangles of the form $l \times \pi$.*

Solution of the plane geometry problem

In this subsection we prove the incommensurability lemma. The proof is contained in problem 7.9.4.

Let a set of rectangles be given. One can get a new set by cuttting one of these rectangles into two new ones. Such the operation is called an *elementary transformation* of the set.

7.9.4. (a) If two sets of rectangles are rectangular scissor-congruent, then one of them can be obtained from the other by a sequence of elementary transformations, transformations inverse to them, and parallel translations.

Let θ and π be incommensurable. Suppose, on the contrary, that from the rectangle $a \times \theta$ we get the rectangle $b \times \pi$ by a sequence of elementary transformations, transformations inverse to them, and parallel translations of parts. Let $\theta, \pi, y_1, y_2, y_3, \ldots, y_N$ be the lengths of the vertical sides of all rectangles that occurred in this sequence of elementary transformations. Denote

$$Y := \{\theta, \pi, y_1, \ldots, y_N\}.$$

(b) There exist numbers $\{y_1', y_2', \ldots, y_n'\} \subset Y$ such that any number $y \in Y$ is uniquely represented in the form

(*) $y = p\theta + q\pi + p_1 y_1' + p_2 y_2' + \cdots + p_n y_n'$,

for some rational numbers p, q, p_1, p_2, $\ldots$, p_n.

Fix a set of such numbers $\theta, \pi, y_1', \ldots, y_n'$ (called the *basis*). For each $y \in Y$ let $f(y) := p$, where p is the coefficient at θ in (*)

If M is a set of $x_1 \times y_1$, $x_2 \times y_2$, $\ldots$, $x_n \times y_n$ rectangles, where $y_i \in Y$, then define

$$J(M) := x_1 f(y_1) + x_2 f(y_2) + \cdots + x_n f(y_n).$$

(c) The value of $J(M)$ does not change under elementary transformations of M.

(d) Prove the incommensurability lemma.

(e) Prove Dehn's theorem: the regular tetrahedron and cube are not scissors-congruent.

This method allows us to establish other interesting facts about cutting of figures.

7.9.5. (a) Prove another theorem due to Dehn: if an $a \times b$ rectangle can be cut into squares, then a/b is rational (this was problem 7.8.21).

(b) Prove that a regular tetrahedron cannot be cut into two or more regular tetrahedra.

Another approach to solving problem 7.9.5(a) is outlined in Section 8 of this chapter; see 7.8.13–7.8.21.

Conclusion: complete invariants

The set of rectangles that correspond to a polytope is called the *Dehn invariant*.[3] Surprisingly, the statement which, in a sense, is reciprocal to the reduction lemma is also true.

If two polytopes have equal volumes and their corresponding sets of rectangles become rectangular scissor-congruent after adding rectangles of the form $l \times \pi$, then the two original polytopes are scissor-congruent.

The rectangular scissor-congruent invariant $J(M)$ of sets of rectangles in the plane which we constructed in problem 7.9.4 is not a complete invariant. Using similar methods, however, a complete invariant can be constructed (it is called the *Kenyon invariant*; see [**Ken**]).

Suggestions, solutions, and answers

7.9.1. (a) Cut the original parallelepiped into two $1 \times 2 \times 2$ parallelepipeds.

(b) The vertices of the required pyramids are in the center of the cube; the bases are the faces of the cube.

(c) *Geometric solution.* The cube $ABCDA'B'C'D'$ is cut into 6 tetrahedra $AC'BB'$, $AC'B'A'$, $AC'A'D'$, $AC'D'D$, $AC'DC$, $AC'CB$ by six planes passing through a pair of opposing vertices A, C' of the cube and one of the remaining vertices. The congruence of these tetrahedra follow from symmetry considerations (e.g., tetrahedron $AC'BB'$ maps to tetrahedron $AC'A'D'$ when you turn the cube by $120°$ about the line AC').

Algebraic solution. The cube $0 \le x, y, z \le 1$ can be cut into 6 tetrahedrons:

$$0 \le x \le y \le z \le 1, \quad 0 \le x \le z \le y \le 1, \quad 0 \le y \le x \le z \le 1,$$
$$0 \le y \le z \le x \le 1, \quad 0 \le z \le x \le y \le 1, \quad 0 \le z \le y \le x \le 1.$$

In fact, we described the same cutting in two different ways.

---

[3]This definition is equivalent to the generally accepted algebraic definition in [**Fuks**].

7.9.3. (a) *Solution.* Let e_i be the edge of some polytope M_j lying on the edge e. Consider the cylinder C with axis e and radius 1 (its bases pass through the endpoints of e). The dihedral angle at e cuts out a "ribbon" L on the surface of the cylinder of length l (along the direction of the cylinder axis) and angular "width" α (in the lateral surface of the cylinder). On the surface of a smaller cylinder C_i with axis e_i and radius 1, the dihedral angle of the edge e_i of M_j cuts out a ribbon L_i of length l_i and width α_i. Since the polyhedra in the cutting do not intersect and cover the entire polytope M, the ribbon L is sliced into smaller ribbons $L_1, L_2, \ldots, L_n$. It remains to map the cylinder to the plane, i.e., to establish a natural correspondence between the points of the ribbon L and the rectangle $l \times \alpha$, and we get the cutting of this rectangle into rectangles $l_i \times \alpha_i$.

(b) *Solution.* Any point on a line ℓ belonging to the polytope M is either an interior point of some M_i or lies on the boundary of several polytopes in the cutting. Let $e_1, e_2, \ldots, e_n$ denote all the edges of the polytopes of the partition that lie on ℓ (denote their lengths by $l_1, l_2, \ldots, l_n$). The union of all the edges $c_1, e_2, \ldots, e_n$ forms a family of line segments on ℓ. Without loss of generality we can assume that $e_1, e_2, \ldots, e_s$ are all the edges lying on one of the segments I of this family. Let $f_1, f_2, \ldots, f_m$ be all possible intersections of this segment I with the interiors of the faces of the polytopes M_j.

Let us prove that the set of $e_1 \times \alpha_1$, $e_2 \times \alpha_2$, $\ldots$, $e_s \times \alpha_s$ rectangles is rectangular scissor-congruent to some rectangle of the form $l \times \pi$. Then, performing the same operation for all rectangles of width π obtained from all the segments in our family, we obtain the required family.

Let C, C_i, C_j be cylinders of radius 1 with axes I, e_i, f_j, respectively. The dihedral angle at the edge e_i in the corresponding polytope cuts out an $l_i \times \alpha_i$ ribbon on the surface of the cylinder C_i. The plane of the face of M_j that contains the segment f_j, slices two $f_j \times \pi$ ribbons on C_j; take the ribbon lying in the same half-space as M_j. Since the polytopes are disjoint and cover M, the cylinder C is cut into $l_i \times \alpha_i$ and $f_j \times \pi$ ribbons.

Extend all the cuts perpendicular to the axis of the cylinder, cutting C into rings. Throw out from all the rings the ribbons of width π (parts of the "extra" ribbons of length f_j), and cut each remaining ring into two ribbons of width π. From these ribbons, we assemble a ribbon of width π. It is a rectangle of width π cut into $l_1 \times \alpha_1$, $l_2 \times \alpha_2$, $\ldots$, $l_s \times \alpha_s$ rectangles obtained by vertical and horizontal cuts and parallel translations.

(c) *Solution.* Suppose M and M' are scissor-congruent. Let $M_1, M_2, \ldots, M_n$ be a set of convex polytopes from which one can assemble both M and M'. By problems 7.9.3(a) and (b), the set of rectangles corresponding to the polyhedron M, combined with some set of rectangles of width π, is rectangular scissor-congruent to the union of the sets of rectangles corresponding to the polytopes $M_1, M_2, \ldots, M_n$. The same is true for M'. However, it is obvious that the relation of rectangular scissor-congruence is transitive and symmetrical. This means that the sets of rectangles corresponding to the polytopes M and M' become equal after adding to them suitable rectangles of width π, as required.

(d) *Solution.* Consider the tetrahedron $ABCD$. Let M be the midpoint of CD. Since the segments AM and BM are perpendicular to CD, the angle AMB is the dihedral angle at the edge CD of the tetrahedron. Let the edge length of a tetrahedron be a. Since the faces of the tetrahedron are equilateral triangles, $AM = BM = \frac{\sqrt{3}}{2}a$. The law of cosines applied to triangle AMB yields

$$\cos\theta = \frac{AM^2 + BM^2 - AB^2}{2AM \cdot BM} = \frac{1}{3}.$$

We now prove by induction that $\cos n\theta = a_n/3^n$, where a_n is an integer not divisible by 3. The base cases $n = 0$ and $n = 1$ are obvious.

Next, let $n \geq 1$. By the formula for the sum of cosines we have

$$\cos(n+1)\theta + \cos(n-1)\theta = 2\cos n\theta \cos\theta,$$

so that

$$\cos(n+1)\theta = 2\cos n\theta \cos\theta - \cos(n-1)\theta = \frac{2a_n - 3a_{n-1}}{3^{n+1}}.$$

By the induction hypothesis, a_n is not divisible by 3. Therefore, $a_{n+1} = 2a_n - 3a_{n-1}$ is not divisible by 3.

Therefore, $\cos n\theta \neq 1$ for any integer n, so $n\theta \neq 2\pi m$ for any integers n and m, which implies that $\theta \neq \frac{m}{n} \cdot 2\pi$. The statement is proved.

7.9.4. (a) *Solution.* Let the first set be cut into rectangles and then the pieces be moved to assemble the second set. Make additional cuts: continue all vertical cuts in the cutting of the first set and all horizontal cuts in the cutting of the second set. The resulting cutting can be performed by a sequence of elementary transformations and inverses of them: first, cut the first set sequentially along all vertical cuts, and then cut each of the resulting vertical strips by horizontal cuts. Now collect the horizontal strips of the partition of the second set, and then assemble them into the rectangles of the second set.

(b) *Suggestion.* See section "A quite good proof of Lemma 2" in [**Fuks**] or the solution to problem 7.8.18.

(c) *Solution.* Cut an $x \times y$ rectangle into two rectangles $x_1 \times y$, $x_2 \times y$ or $x \times y_1, x \times y_2$.

The case where the cut is vertical is obvious; indeed, the invariant $J(M)$ has changed by

$$x_1 f(y) + x_2 f(y) - x f(y) = 0;$$

i.e., it does not change.

Suppose the cut is horizontal. Then the invariant changes by $xf(y_1) + xf(y_2) - xf(y)$. Let us prove that this value is equal to zero. Let

$$y_1 = f(y_1)\theta + q_1\pi + \mu_1 y_1' + \mu_2 y_2' + \cdots + \mu_n y_n',$$
$$y_2 = f(y_2)\theta + q_2\pi + \xi_1 y_1' + \xi_2 y_2' + \cdots + \xi_n y_n'.$$

Then

$$y = y_1 + y_2 = (f(y_1) + f(y_2))\theta + (q_1 + q_2)\pi$$
$$+ (\mu_1 + \xi_1)y_1' + (\mu_2 + \xi_2)y_2' + \cdots + (\mu_n + \xi_n)y_n'.$$

Thus $f(y) = f(y_1) + f(y_2)$, and our invariant has not changed.

(d) *Solution.* Since the invariant $J(M)$ is preserved under elementary transformations, part (a) above implies that the invariants of any two rectangular scissor-congruent sets are equal. However, the invariant for the set $(a \times \theta, l \times \pi)$ is equal to a, and for $b \times \pi$ it is equal to zero. Therefore, these sets are not rectangular scissor-congruent.

(e) *Solution.* Assume that a regular tetrahedron with an edge a and a cube with an edge b are scissor-congruent. Then, by the reduction lemma, the set consisting of six $a \times \theta$ rectangles and one $l_1 \times \pi$ rectangle is rectangular scissor-congruent to the set of eight $b \times \frac{\pi}{2}$ rectangles and one $l_2 \times \pi$ rectangle. However, the first set is rectangular scissor-congruent to the set $6a \times \theta, l_1 \times \pi$, and the second is rectangular scissor-congruent to the set $\left(\frac{b}{2} + l_2\right) \times \pi$. Therefore, the last two sets are rectangular scissor-congruent. However, the incommensurability lemma and problem 7.9.3(d) imply that they are not rectangular scissor-congruent, a contradiction.

7.9.5. (a) *Solution.* If an $a \times b$ rectangle can be cut into squares, then it is rectangular scissor-congruent to a $b \times a$ rectangle (since the square transforms into itself when rotated through $90°$). By the incommensurability lemma, $\frac{a}{b}$ is rational.

(b) *Algebraic solution.* Assume that the tetrahedron with edge of length a can be cut into two or more tetrahedra. Let $a_1, a_2, \ldots, a_n$ be the edges of these tetrahedra. By problems 7.9.3(a) and (b) (similarly to the reduction lemma), the set $6a_1 \times \theta, 6a_2 \times \theta, \ldots, 6a_n \times \theta$ is rectangular scissor-congruent to the set $6a \times \theta, l \times \pi$. By problems 7.9.4(a) and (c) we have

$$a_1 + a_2 + \cdots + a_n = a.$$

The equality of volumes implies

$$a_1^3 + a_2^3 + \cdots + a_n^3 = a^3.$$

Raising the first equality to the third power we get

$$a_1^3 + a_2^3 + \cdots + a_n^3 + A = a^3,$$

where $A > 0$, and we arrive at a contradiction with the second equality.

Geometric solution. Suppose that the tetrahedron can be cut into two or more tetrahedra. The face of the original tetrahedron cannot be a face of any of the tetrahedra into which it is cut. Hence, the edge of one of the smaller tetrahedra entirely lies on the face of the bigger one. In this edge, several dihedral angles equal to θ meet, and their sum is equal to π. This contradicts problem 7.9.3(d).

Bibliography

[PROB1]   Internet project "Problems", http://problems.ru.

[PROB2]   *Problems from "Kvant"*, Kvant (2013), no. 1, 23–31.

[AS]   Titu Andreescu and Mark Saul, *Algebraic inequalities: new vistas*, MSRI Mathematical Circles Library, Mathematical Sciences Research Institute, Berkeley, CA; American Mathematical Society, Providence, RI, 2016. MR3585329

[Ar98]   V. I. Arnold, *Statistics of the first digits of powers of two and redivision of the world*, Kvant (1998), no. 1, 2–4.

[Ar16]   Vladimir Arnold, *Problems for children 5 to 15 years old*, Eur. Math. Soc. Newsl. **98** (2015), 14–20. Excerpt from [MR3409220]. MR3445185

[B]   E. J. Barbeau, *Polynomials*, Springer-Verlag, 2003.

[BMSCS]   A. Belov, I. Mitrofanov, A. Skopenkov, A. Chilikov, and S. Shaposhnikov, *13th Hilbert's problem on superpositions of functions*, https://www.turgor.ru/lktg/2016/5/index.htm.

[Bew]   Jörg Bewersdorff, *Galois theory for beginners: A historical perspective*, translated from the second German (2004) edition by David Kramer, Student Mathematical Library, vol. 35, American Mathematical Society, Providence, RI, 2006, DOI 10.1090/stml/035. MR2251389

[Bol78]   V. Boltyansky, *How often do powers of two begin with 1*, Kvant (1978), no. 5, 2–7.

[Bol56a]   V. G. Boltyansky, *Equal-area and scissors-congruent figures*, Gostekhizdat, Moscow, 1956. (Russian)

[Bug]   V. O. Bugaenko, *Lomonosov's Tournaments mathematical competitions*, Moscow Center for Continuous Mathematical Education, 1998.

[CF]   N. Casey and M. Fellows, This is MEGA-mathematics! Stories and activities for mathematical thinking problem-solving and communication, the Los Alamos workbook, http://www.c3.lanl.gov/~captors/mega-math, 1993.

[CRM]   M. Cencelj, D. Repovs, and M. Skopenkov, *A short proof of the twelve points theorem*, Math. Notes **77** (2005), no. 1, 108–111.

[Chen]   Evan Chen, *Euclidean geometry in mathematical olympiads*, AMS/MAA Problem Books Series, Mathematical Association of America, Washington, DC, 2016. MR3467691

[ChS]   A. Chhartishvili and E. Shikin, *Dynamic games of simple search*, Kvant (1996), no. 1, 6–12.

[CoxGr]   H. S. M. Coxeter and S. L. Greitzer, *Geometry revisited*, New Mathematical Library, vol. 19, Random House, Inc., New York, 1967. MR3155265

[FBKYa1]   Roman Fedorov, Alexei Belov, Alexander Kovaldzhi, and Ivan Yashchenko, *Moscow Mathematical Olympiads, 1993–1999*, Amer. Math. Soc., Providence, RI, 2011.

[FBKYa2]   Roman Fedorov, Alexei Belov, Alexander Kovaldzhi, and Ivan Yashchenko,
           Moscow Mathematical Olympiads, 2000–2005. Amer. Math. Soc., Providence,
           RI, 2011.

[FGI]   Dmitri Fomin, Sergey Genkin, and Ilia Itenberg, *Mathematical circles (Russian
           experience)*, translated from the Russian and with a foreword by Mark Saul,
           Mathematical World, vol. 7, American Mathematical Society, Providence, RI,
           1996, DOI 10.1090/mawrld/007. MR1400887

[FKh]   Dmitry Fomin and Alexey Kirichenko, *Leningrad Mathematical Olympiads
           1987–1991*, Contests in Mathematics, vol. 1, translated from the Russian by
           Fomin and Kirichenko, with a foreword by Mark Saul and a preface by Stanley
           Rabinowitz, MathPro Press, Westford, MA, 1994. MR1374787

[F]   Ronald M. Foster, *The average impedance of an electrical network*, Reissner
           Anniversary Volume, Contributions to Applied Mechanics, J. W. Edwards,
           Ann Arbor, MI, 1949, pp. 333–340. MR29773

[Fuks]   D. Fuchs, *Is it possible to make a cube from a tetrahedron?*, Kvant (1990), no.
           11, 2–11.

[Fut]   A. Futer, *Signals, graphs, and kings on a torus*, Kvant (1977), no. 7, 14–19.

[G]   M. Gardner, *Squaring a square*, Math puzzles and entertainments, Mir, 1999.

[GK]   S. Genkin and L. Kurlyandchik, *Numeric constructions*, Kvant (1990), no. 9,
           58–61.

[Gik]   E. Gik. *Crosses and noughts*, Kvant (1994), no. 1, 53–54.

[GKP]   Ronald L. Graham, Donald E. Knuth, and Oren Patashnik, *Concrete mathe-
           matics: A foundation for computer science*, 2nd ed., Addison-Wesley Publishing
           Company, Reading, MA, 1994. MR1397498

[GDI2]   A. A. Glibichuk, A. B. Dainyak, D. G. Ilyinsky, A. B. Kupavsky, A. M. Raig-
           orodsky, A. B. Skopenkov, and A. A. Chernov, *Elements of discrete mathe-
           matics in problem*, MCCMO, Moscow, 2016. Abridged version: http://www.
           mccme.ru/circles/oim/discrbook.pdf.

[GZ]   S. M. Gusejn-Zade, *Picky bride*, MCCME, Moscow, 2003 (Russian).

[G]   V. Gutenmakher, *Linear equation systems*, Kvant (1984), no. 1, 24–29.

[K-BS]   A. Kanel-Belov and M. Sapir, *"And the wind returns again to its circuits", or
           periodicity in math*, Kvant **56** (1990), no. 4, 6–10.

[K-BK92]   A. Ya. Kanel-Belov and A. K. Kovaldzhy, *Triangles and catastrophes*, Kvant
           (1992), no. 11, 42–50.

[K-BK08]   A. Ya. Kanel-Belov and A. K. Kovaldzhy, *How to solve non-standard problems:
           fourth edition*, Moscow Center for Continuous Mathematical Education, 2008.

[KG]   A. E. Karpov and E. Ya. Gik, *Chess Kaleidoscope*, "Kvant" library, Pergamon
           Press, 1981.

[K]   Ya. Karpov, *Optimal postcode encoding*, Kvant (1987), no. 11, 19–20.

[KKPV]   Kiran S. Kedlaya, Bjorn Poonen, and Ravi Vakil, *The William Lowell Put-
           nam Mathematical Competition, 1985–2000, Problems, solutions, and commen-
           tary*, AMS/MAA Problem Books Series, Mathematical Association of America,
           Washington, DC, 2002. MR1933844

[Ken]   Richard Kenyon, *Tilings and discrete Dirichlet problems*, Israel J. Math. **105**
           (1998), 61–84, DOI 10.1007/BF02780322. MR1639727

[Kh]   A. G. Khovanskiĭ, *Newton polytopes, curves on toric surfaces, and inversion of
           Weil's theorem* (Russian), Uspekhi Mat. Nauk **52** (1997), no. 6(318), 113–142,
           DOI 10.1070/RM1997v052n06ABEH002156; English transl., Russian Math.
           Surveys **52** (1997), no. 6, 1251–1279. MR1611333

[KZhP]   A. N. Kolmogorov, I. G. Zhurbenko, and A. V. Prokhorov, *Introduction to
           Probability Theory*, Number 135 in "Kvant" library, Moscow Center for Con-
           tinuous Mathematical Education, 2015.

[Kor]   B. Kordemsky, *To the Repunits family for an hour*, Kvant (1997), no. 5, 28–29.

[Kru]   R. Krutovsky, *On graphs of a given diameter without small cycles*, `http://` `www.mccme.ru/circles/oim/mmks/works2013/krutowski2.pdf`.

[Kur1]   Jozsef Kurschak, *Hungarian Problem Book I*, The Mathematical Association of America, 1963.

[Kur2]   Jozsef Kurschak, *Hungarian Problem Book II*, The Mathematical Association of America, 1963.

[Kush1]   A. Kushnirenko, *Integer points in polygons and polyhedrons*, Kvant (1977), no. 4, 13–20.

[Kush2]   A. Kushnirenko, *Newton polygon*, Kvant (1977), no. 6, 19–24.

[Leb]   Henri Lebesgue, *Measure and the integral*, edited with a biographical essay by Kenneth O. May, Holden-Day, Inc., San Francisco, Calif.-London-Amsterdam, 1966. MR201592

[Liu]   Andy Liu, editor, *Hungarian Problem Book III*, The Mathematical Association of America, 2001.

[LT]   S. Lvovsky and A. Toom, *Possible and impossible*, Kvant (1989), no. 1, 52–55.

[MS]   A. Matulis and A. Savukinas, *Put queen in the corner, tsyanshidzi and Fibonacci numbers*, Kvant **29** (1984), no. 7, 18–21.

[MedSha] L. Eh. Mednikov and A. V. Shapovalov, *Tournament of Towns, The world of Mathematics in Problems*, MCMME, Moscow, 2012.

[Mos]   Frederick Mosteller, *Fifty challenging problems in probability with solutions*, reprint of the 1965 original, Dover Publications, Inc., New York, 1987. MR896688

[Or76]   A. Orlov, *Item search*, Kvant (1976), no. 7, 55–57.

[Or77]   A. Orlov, *Put on minus!*, Kvant (1977), no. 3, 41–45.

[Pev]   P. Pevzner, *The best bet for simpletons*, Kvant (1987), no. 5, 4–8.

[Pl]   Plato, *Phaedo*, Kindle Edition, Amazon Digital Services, 2012.

[PS]   G. Pólya and G. Szegő, *Problems and Theorems in Analysis*, Springer-Verlag, 2004.

[PR-V]   Bjorn Poonen and Fernando Rodriguez-Villegas, *Lattice polygons and the number 12*, Amer. Math. Monthly **107** (2000), no. 3, 238–250, DOI 10.2307/2589316. MR1742122

[PraSko] M. Prasolov and M. Skopenkov, *Tiling by rectangles and alternating current*, J. Combin. Theory Ser. A **118** (2011), no. 3, 920–937, DOI 10.1016/j.jcta.2010.11.012. MR2763046

[Rai]   A. M. Raigorodsky, *Danzer–Grunbaum acute-angled triangles*, Number 36 in library "Mathematical enlightenment", Moscow Center for Continuous Mathematical Education, 2009.

[Ros]   A. Rosental, *The extreme principle*, Kvant (1988), no. 9, 53–57.

[SA]   L. Sadovsky and M. Arshinov, *Binary coding*, Kvant (1979), no. 7, 14–20.

[S72]   A. Savin, *Encircling of landing force*, Kvant (1972), no. 3, 24–27.

[S87]   A. Savin, *Dissection problems*, Kvant (1987), no. 7, 44–47.

[ShS]   G. Shabat and A. Sgibnev, *Gluing of polygons*, Kvant (2011), no. 3, 17–22.

[Sha08]   A. V. Shapovalov, *Principle of bottlenecks: second edition*, Moscow Center for Continuous Mathematical Education, 2008.

[Sha14]   A. V. Shapovalov, *How to build an example*, School Math Circles, Moscow Center for Continuous Mathematical Education, 2014.

[Sha15]   A. V. Shapovalov, *Mathematical constructions: from huts to palaces*, School Math Circles, Moscow Center for Continuous Mathematical Education, 2015.

[Sh]   F. Sharov, *Cutting a rectangle into rectangles with given aspect ratios*, Matematicheskoye Prosveshcheniye (Mathematical Enlightenment) **20** (2016), 200–214.

[Shes]   G. Shestopal, *How to detect a fake coin*, Kvant (1979), no. 10, 21–25.

[SC]   D. O. Shklyarsky and N. N. Chentzov, editors, *The USSR Olympiad Problem Book: Selected Problems and Theorems of Elementary Mathematics*, Dover, 1993.

[Sko]   Arkadiy Skopenkov, *Mathematics via problems. Part 1. Algebra*, translated from the Russian original by Paul Zeitz and Sergei G. Shubin, with a foreword by Paul Zeitz, MSRI Mathematical Circles Library, vol. 25, Mathematical Sciences Research Institute, Berkeley, CA; American Mathematical Society, Providence, RI, [2021] ©2021. MR4242842

[SMD]   M. Skopenkov, O. Malinovskaya, and S. Dorichenko, *Compose a square*, Kvant (2015), no. 2, 6–11.

[SPD]   M. Skopenkov, M. Prasolov, and S. Dorichenko, *Dissection of a metal rectangle*, Kvant (2011), no. 3, 10–16.

[SSU]   M. Skopenkov, V. Smykalov, and A. Ustinov, *Random walks and electrical circuits*, Matematicheskoye Prosveshcheniye (Mathematical Enlightenment) **16** (2012), 25–47.

[Sob]   S. Sobolev, *About random walks*, Kvant (1987), no. 5, 45–49.

[Soi]   A. Soifer, *Checkered boards and polyominos*, Kvant (1972), no. 11, 2–10.

[St]   J. Michael Steele, *The Cauchy-Schwarz master class: An introduction to the art of mathematical inequalities*, AMS/MAA Problem Books Series, Mathematical Association of America, Washington, DC; Cambridge University Press, Cambridge, 2004, DOI 10.1017/CBO9780511817106. MR2062704

[Tab]   S. Tabachnikov, *Geometry of equations*, Kvant (1998), no. 10, 10–16.

[TF]   S. Tabachnikov and D. Fuchs, *Impossible tilings*, Kvant (2011), no. 2.

[Tm]   Z. Tmeladze, *Game theory*, Kvant (1977), no. 8, 27–33.

[Vas91]   N. Vasiliev, *Geometric probabilities*, Kvant (1991), no. 1, 47–53.

[Vas74]   N. B. Vasiliev, *Around Pick's theorem*, Kvant (1974), no. 12, 39–43.

[Ver]   B. Vertgejm, *Games with quadratic functions*, Kvant (1981), no. 11, 6–10.

[Vil71a]   A. Vilenkin, *Problems from "Kvant". Solutions*, Kvant (1971), no. 3, 33–35.

[Vil71b]   N. Vilenkin, *Combinatorics*, Kvant (1971), no. 1, 13–19.

[VINK]   O. Ya Viro, O. A. Ivanov, N. Yu. Netsvetaev, and V. M. Kharlamov, *Elementary Topology*, Moscow Center for Continuous Mathematical Education, 2010.

[Yad]   M. J. Yadrenko, *Pigeonhole principle and its application*, Vischa Shkola, 1985.

[Yag71]   I. Yaglom, *Two games with matches*, Kvant (1971), no. 2, 4–10.

[Yag74]   I. Yaglom *Patches on a caftan*, Kvant (1974), no. 2, 13–21.

[Yag68]   I. M. Yaglom, *How to dissect a square?*, Math. Bibl., Nauka, Moscow, 1968, 112 pp.

[Yu]   A. Yuriev, *Random walks are returning*, Mathematical enlightenment **20** (2016), 243–246.

[Z]   S A. Zaslavsky, *On dissecting a convex polygon*, Kvant (2012), nos. (5–6), 53–55.

[ZS]   Alexey A. Zaslavsky and Mikhail B. Skopenkov, *Mathematics via problems. Part 2. Geometry*, translated from the 2018 Russian original by Paul Zeitz and Sergei G. Shubin, MSRI Mathematical Circles Library, vol. 26, Mathematical Sciences Research Institute, Berkeley, CA; American Mathematical Society, Providence, RI, [2021] ©2021. MR4506521

[ZSS]   A. Zaslavsky, A. Skopenkov, and M. Skopenkov, editors, *Elements of mathematics in problems: from Olympiads and circles to the profession itself*, Moscow Center for Continuous Mathematical Education, 2017.

Index

PUBLISHED TITLES IN THIS SERIES